MW01609058

Anunnaki Series

A set of 3 Books

**The Anunnaki and Ulema-Anunnaki Vault of Forbidden
Knowledge and the Universe's Greatest Secrets. Book 3**

A set of 3 books

The Anunnaki and Ulema-Anunnaki Vault of Forbidden Knowledge and the Universe's Greatest Secrets.

Book 3

A set of 3 books

Maximillien de Lafayette

*** *** ***

Contributions by
Ulema Mordachai ben Zvi
Ulema Kira Yerma
Ulema Ramash Govinda
Master Li

Times Square Press. Elite Associates International

Jamiyat Ramadosh Al-Ulema Al-Anunnaki
New York California London Paris Tokyo

*** *** ***

2010

Acknowledgment and Gratitude

I am deeply grateful to the Honorable Anunnaki-Ulema who have generously contributed to this book:
Contributors:
Master Li
Ulema Sharif Al Mutawalli
Ulema Mordachai Ben Zvi
Ulema Sadik Bin Jaafar Al Kamali
Ulema F. Oppenheimer
Ulema A. Berkof
Ulema Ramash Govinda
Ulema Tabeth Al-Baydani
Ulema Shaul Sorenztein
Sinhar Ambar Anati
Ulema F. Tayara
Ulema Mirach Faridi Beraz
For without their guidance and contributions, this book would have remained stacks of papers in my drawer.

⌘ ⌘ ⌘

Table of Contents

Table of Contents (Book 3)

110. Anunnaki-Ulema...143

111. An-Hayya'h, "A-haYA", "Aelef-hayat"...155

112. Apindugari: The Anunnaki's Wedding...163

117. Humans' Link To The Beginning Of Everything, The Anunnaki, and "God"...217

118. The Anunnaki Christian Saint: St. Tekla...237

119. Jesus did not die on the Cross...257

Ambar Anati goes back in time, and revisits Mary Magdalene and sees Jesus Christ alive in Marseille, France...257

---

82. Images in Another Dimension
"Lariba"
"Marda-iruch"
⌘⌘⌘

I. Definition
II. Ulema Tabet Al Ansari bin Koufia explains

I. Definition:
The animated pictures and images of usually immobile objects, you see when you enter another dimension.

II. Ulema Tabet Al Ansari bin Koufia explains:
The Honorable Master said verbatim:

- In some dimensions, objects that usually lack motion on Earth, move freely on their own, as if they have a motion motor inside them.
- Many of the objects you see here on Earth do not move. Objects like stones, chairs, mountains, etc.
- In an extra dimension, all objects extend the place they occupy.
- They are not limited by their superficies and borders.
- On Earth, we understand and define all sorts of things by calculating time and space.
- Usually time is calculated by measuring distances.
- Humans understand distances by measuring time.
- And this, is not totally correct. Especially when we try to define speed.
- In other spheres of existence, these objects have an extra dimension.
- It is a very complicated matter for the human mind.

- These objects, in addition to their structural properties have images of their own.
- These images are part of the physical place they usually occupy.
- In other spheres, you can move these huge objects, like mountains for instance, by moving their images.
- The Anunnaki can move extremely heavy objects, like for example, 1,500 tons stones or temples without touching them physically, or using any physical means, such as cranes or winches.
- They do this by creating an animated picture of the heavy object, transposing the image on a matrix, and moving the matrix from one place to another.
- It is not a teleportation.
- It is rather a transposition of the physical properties of an object.
- The Anunnaki can even move a whole city from one country to another country, simply by moving the holographic image of the city.
- The Anunnaki can easily create a holographic image of anything that exists in the world.
- So when I said "Objects move freely on their own, as if they have a motion motor inside them", I meant, that an image of these objects has been created, and transported to another spot.
- This image once transported to another spot becomes reality.
- It is like duplicating an object and moving it somewhere else.
- Now, you might ask: "Then what happens to the real object the Anunnaki duplicated as an image and moved it somewhere else?
- Is it still there, or just vanished?"
- It did not vanish.
- It has been transported.
- And now, the object has become an image.
- But this image is real, and can take a physical form.
- Your mind cannot understand this phenomenon, because you are still thinking about matter, time, space and distance.

83. Higher Level of Knowledge "Malak"
⌘ ⌘ ⌘

I. Definition
II. Children of the heaven
III. Their leaders are

_____

I. Definition:
Hebrew, Aramaic, Ulemite, and Arabic word for an angel.
It derived from the Anunnaki's terminology of Malka, which means kingdom or a higher level of knowledge and mental development.

From Malka, derived:
- **a**- The Hebrew word Malkoth, which means the kingdom of God;
- **b**-The Arabic word Malakoot, which means the kingdom or god, and divine rule.
- **c**-The Proto-Aramaic and Aramaic word Malakut, which means the kingdom of God;
- **d**- The Syriac word Malakout, which means heaven, and the heavenly kingdom;
- **e**-Malakout, which means the Divine One in the Brahman literature.

Malak in Anunnaki's literature means a messenger.
And from Malak, derived:
- **a**-The Arabic word Malaak, which means an angel; Malaa'ikah in plural.
- **b**-The Hebrew word Malak, which means an angel. Malakim in plural.
- **c**-The Ethiopic word Malak, which means angel. Mala'ikt in Plural

II. Children of the heaven:

Some theologians, including the early Hebraic scholars, and Kabalists called the Malakim "Malaa'ikah", the children of the heaven.

III. Their leaders are:

- Samlazaz,
- Araklba,
- Rameel,
- Kokablel,
- Tamlel,
- Ramlel,
- Danel,
- Ezeqeel,
- Baraqijal,
- Asael,
- Armaros,
- Batarel,
- Ananel,
- Zaqel,
- Samsapeel,
- Satarel,
- Turel,
- Jomjael,
- Sariel.

Worth mentioning here that:

- **a**-Ezeqeel derived from the Anunnaki's word Ez-ikil.
- **b**- Sariel derived from the Anunnaki's word Sarim.
- **c**- Ananel derived from the Anunnaki's word Ana. Il.

*** *** ***

84. Humans' Early Species
⌘ ⌘ ⌘

1. The Akama-ra
2. The Bashar; the 36 (Some say 46) different human and quasi-human species who lived on Earth
3. Some of the early human forms were
a- The Earth-made human creatures
b- The space-made human creatures
c- Metabolism and the oceans-made human creatures

84. Humans' Early Species
⌘ ⌘ ⌘

1. The Akama-ra
2. The Bashar; the 36 (Some say 46) different human and quasi-human species who lived on Earth
3. Some of the early human forms were
a- The Earth-made human creatures
b- The space-made human creatures
c- Metabolism and the oceans-made human creatures

---

1. The Akama-ra:

Akarama-ra is an Anunnaki word/term in the Ana'kh language. According to the Ulema, the Akama-ra were the first beings who were allowed by the Anunnaki (Enki and Inanna) to date the "Women of Light" who were quarantined on earth by the Anunnaki. Akama-ra were genetically created by the Anunnaki on Ashtari (Nibiru) and were transported to planet earth on Anunnaki's spaceships, called Merkabah. The Ulema said that the Akama-ra were the first beings who were allowed by Enki and Inanna to date the "Women of Light" who were quarantined on Earth by the Anunnaki. Akama-ra were genetically created by the Anunnaki on Ashtari (Nibiru) and were transported to planet earth on Anunnaki's spaceships, called Merkabah.

2. The Bashar; the 36 (Some say 46) different human and quasi-human species who lived on Earth:

Bashar is a name given by the extraterrestrial Anunnaki to the first human beings, who appeared on planet earth in a multitude of forms and shapes. Excerpt from their Kira'at (Reading): "On planet earth existed many different human races for millions of years.

Some are known to us, while many others are totally unknown because they have vanished without leaving a trace.

The truth is that they have left many traces, but we did not discover them yet. In the near future, we will discover some of their remains, and a new chapter on the history of mankind will be written."

However, in 2003, skeletons of four vanished early forms of humans who did not look like humans were discovered by English archeologists and anthropologists, but were shrouded in secrecy, and their discoveries were never made public for many reasons.

Two leading and extremely powerful Catholic theologians were behind the cover-up. Some of those early quasi-human forms were 10 feet in height, and others less than 2 feet, and looked like hobbits. Those species were created by various extraterrestrial races. The Anunnaki did not take part in the creation process of these very tall and very small quasi-humans.

The extraterrestrials created them on planet earth. But there are other early human beings who were created in space, and on other planets, and like the very small and very tall species, they were not part of the evolutionary process of the modern human beings.

In total, 36 (Some say 46) different human and quasi-human species lived on planet earth in many regions of the globe. And none of them were created by the God we know and worship. After all, they did not look like humans, and if we have to believe that humans were created in the image of God, as Judaism, Christianity and Islam tell us, then, most certainly those early 36 different species who looked like ferocious beasts, were not made in the image of God. Because they were created in many regions of planet earth, and interbred around the globe, new horrifying species populated the Earth.

So the out-of-Africa theory is entirely wrong. Humanity did neither start in, nor expand from Africa. All those species died out after a very short lifespan, because they did not have in their brains the Anunnaki's Conduit.

3. Some of the early human forms were:
a- The Earth-made human creatures:
The Early quasi-human beings were created from a single cell. The Ulema said that the Anunnaki's Matrix mentioned earth-

made quasi human creatures created from one cell. Those creatures were an organism that looked like bacteria. And the bacteria were in fact, the first life-form on planet earth. Those creatures appeared as tiny mini-microscopic organisms that used chemical energy. They did not need the sun to grow and develop via photosynthesis.

In fact, those early quasi-human forms existed long before photosynthesis developed on planet earth. This explains the reason of their deformity, and horrific dysmorphic features (Dysmorphism).

Those creatures were Earth-made, and first appeared in Australia, Brazil, and Madagascar, not in Africa. They were not created by God.

b- The space-made human creatures:
Some of the very early human races were originally created in space. Their life started within the clay found inside comets. The amazing aspect of this explanation is the fact that most recently, leading scientists and university professors in the United Kingdom have stated, that in fact, the inside of comets contains gluey material identical to heated clay. Once this comet's clay is mixed with earth's water, it creates cells, molecules and membranes. In other words: LIFE.

Authors' note: Recent probes inside comets show it is overwhelmingly likely that life began in space, according to a new research paper by Cardiff scientists.

Professor Chandra Wickramasinghe and colleagues at the Cardiff Centre for Astrobiology have long argued the case for panspermia – the theory that life began inside comets and then spread to habitable planets across the galaxy.

A BBC **Horizon** documentary traced the development of the theory. Now the team claims that findings from space probes sent to investigate passing comets reveal how the first organisms could have started.

c- Metabolism and the oceans-made human creatures:
The Anunnaki's Matrix also revealed the existence and origin of early human like creatures who lived at the bottom of the oceans. The Ulema explained that one of the earliest life-forms on planet earth began at the very bottom of the oceans, where metabolism

originated. Metabolism created an early human like form. Those creatures had a human skull, two eyes without retina, two legs and four long arms, but no nose, no ears, and no hair on their bodies.

They were called the "Basharma'h". (Bashar means human race, and Ma'h means water. Later on in history, these two words were added to several ancient Near and Middle East languages:

a-Mem in Hebrew,
b-Mayim in ancient Aramaic;
c-Ma' or Maay in Arabic, etc...

*** *** ***

85. Anunnaki's Map of the After-life and your Enlightenment on Earth
⌘⌘⌘

Kira'at of Ulema Li:
"Being part of the Anunnaki's tradition, and walking their path, mean enlightenment at two levels: Mental and ethical..." said the Honorable Master Li.

He added, verbatim:
- The mental part develops in you, extraordinary faculties and abilities.
- The ethical part acquaints you with a cosmic truth; a permanent truth that opens your eyes on a way of life, and a code that guarantee access to the metaphysical knowledge.
- The metaphysical knowledge makes you understand what place you are currently occupying now on the landscape of the universe, what is was written on your life page when you were born, even before you were born on Earth, and above all, what your destiny and place are going to be in the after-life.
- I said "even before you were born on Earth", because you have more than one physical existence.
- This means that you co-exist, you have co-existed, and you shall co-exist simultaneously in different planes, and in multiple dimensions.
- Yes, it is possible to learn about what is going to happen to you, when you leave this Earth, and after you enter the next dimension.
- The Mounawariin (The enlightened ones) knew upfront what is going to happen to them after death. And they have learned the Rou'ya (Visions and Scriptures) that will guide them in their passage to the other life.

- In fact, they began to prepare themselves, - right here on Earth – because once you enter a dimension that exists in the after-life, you will become confused and disoriented unless you learn the Rou'ya. Can you learn the Rou'ya? Of course you can, because the Rou'ya is part of your study.
- It is like visiting a new country, a very big country, where streets are very large and very long and lead to different places, many bus stations, strange shops, all the signs in the streets and on the shops' windows are written in a foreign language you don't understand. You look around you, and everything appears strange and stranger. You can't ask for directions, because the people there don't speak a different language. You are totally lost.
- The same thing will happen to you when you enter the dimension of the after-life.
- In that dimension, people speak different languages.
- The streets look different because their layout is so different. The streets lead to multiple destinations, some are physical, others are mental.
- By mental I mean that streets change according to your personal feelings, comprehension, and what you are looking for, or looking at.
- And things speed up differently on other dimensions, because time and space cease to exist.
- Everything you see on Earth, everything you assimilate on Earth is seen and understood according to your understanding and measurement of time on Earth.
- And humans understand time by measuring distance(s).
- In another dimension or layer(s) of the universe, you see people walking in the streets, and you begin to ask yourself, why are they walking like this or like that? Their speed will confuse you.
- In fact, people in that dimension don't walk normally. Some float. In some parts of that dimension, there is no gravity, the kind of gravity we feel and understand on Earth.
- Instead of walking fast to arrive early to your chosen destination, you will be able to bring that destination to you.
- You make distance work for you.

- In fact, you can stay put. You don't have to walk fast at all, because the lack of time will shorten distances. And this will have a major impact on your mind.
- In other parts, the things you see are reflected, just like what normally you see in a mirror.
- Thus, if you are not adequately prepared to enter a no-time/no-space zone, you will be totally lost.
- Learning the Anunnaki's metaphysical truth on Earth allows you to walk freely in the other world.
- Also it allows you to get acquainted with the languages of the other world.
- And the most important thing about all this, is the fact that the metaphysical truth will help you locate and find your loved ones, friends, and even your pets in the next dimension.
- Is the Anunnaki's metaphysical truth similar to the "Third Eye"?
- No, it is not. The Third Eye opens your eyes on the reality and illusion of this world.
- The Third Eye is limited to our physical world, here on Earth. Its powers and wisdom do not transcend the physical perimeter of planet Earth.
- The Anunnaki's metaphysical truth is like a map of a new country you wish to visit for the first time. Even though, you do not speak the foreign language of that country, you will not find any difficulty finding your way around or talking to the people of that dimension in their native tongue, because the map, first shows you all the locations you want to visit, and secondly, it translates the foreign language for you.
- So, you point at one place on the map, and you will immediately understand what people are talking about, you will be able to read all the signs in that foreign language, because the map will serve as a guide, as a translator, and as a dear companion.
- You might ask now, where do I find this map? Is it real, or just the product of my imagination? Is it physical or mental?
- The answer is very simple. It is real and physical. Many Masters saw it here on Earth. And reading the map here

37

on Earth is a major part of preparing yourself to enter a different dimension after you die.

- But only those who are pure in heart, and wish to constantly progress and elevate themselves to a higher level in the after-life will have access to the map. By the way, we call the map in Ana'kh "Falak-Kharta". Falak means the universe, and Kharta means a map.
- Probably you are asking yourself now, or possibly, one day you will ask me: "Master, if the Anunnaki's map of the after-life really works in the after-life, then, why can't I use it here on Earth to make my life a little bit better, to locate friends I have not seen for many years, and I don't know where they are now?
- Why can't I use the Anunnaki's map here on Earth to learn all the different languages on Earth? Why I have to die first to learn all this? And why all these subjects are so mysterious and no direct and simple answers are given to me?
- Well, I have heard all these questions many many times before. When I was a student like you, don't you think I have asked my teachers the very same questions? Of course I did. And do you know what my teachers have told me?
- I will tell you what they have said to me.
- They explained to all the students, that in order to find the Anunnaki's map, and to learn how to use the map, each one of us must become familiar with:
- a- Abgaru
- b- Afik-r'Tanawar
- c- Eido-Rah
- d- Gomari
- Once you have learned a, b, c, d, the Anunnaki's map will reveal itself to you.

*** *** ***

86. Anunnaki Spaceship "Markaba" "Merkabah"
⌘ ⌘ ⌘

I. Definition
II. Etymology

I. Definition:

An Akkadian/Sumerian/Old Babylonian term for a spaceship usually associated with the Anunnaki's space craft. It was mentioned in the Babylonian clay tablets, as well as in the Book of Ramadosh, and the Ulemite literature.

II. Etymology:

- **a-** Markaba in Arabic: The Arabic word "Markab" means a boat. (Plural: Maraakeb.) And the person who rides the Markaba and the Markab is called "Rakeb." The Verb is "Yakkab."
- **b-** Mercavah or Mercabah in Hebrew, and it means a chariot.
- **c-** The Kabalists say that the Supreme after he had established the Ten Sephiroth, he used them as a chariot or throne of glory to descend the souls of men.
- **d-** Worth mentioning here that the word Markaba does not mean UFO and/or a spacecraft in Arabic. The correct name for UFO is "Souhoun Ta-irah" (Flying Saucers).

*** *** ***

87. Miraya
⌘ ⌘ ⌘

Definition:

Miraya is a galactic mirror-monitor, and a sophisticated communication device used by the Anunnaki to view the operations of the Greys.

Miraya enables them to track and watch the activities of the Greys and their allies in their genetic laboratories. Coincidently, the word "Miraya" has been used by the United States military, and the Sudanese authorities in the northern Bahr el-Ghazal State, when they first launched their radio FM station in the area of Aweil; a serious broadcasting-communication operation financed by the United Nations Mission in Sudan (UNMIS).

Miraya is also a Hindu baby name and means Lord Krishna's devotee. But the most striking revelation and linguistic similarity come from the Arabic language, for Miraya means in literary Arabic "Mirror". In Ana'kh (Anunnaki language), Miraya is also a mirror.

***   ***   ***

88. Muluk-Taoos
⌘ ⌘ ⌘

I. Definition
II. Etymology
III. In Ana'kh/Ulemite literature

I. Definition:

Muluk-Taoos is the deity worshipped by the Yezidis, a sect in Persia, called by Christian theology the "devil worshippers", because they worshiped Satan or Lucifer disguised as a peacock. While in fact, in ancient Farsi and archaic Arabic, and in the faith of the Yezidis, "Peacock" does not mean or represent Satan, or the devil.

It is simply the symbol of the "Hundred-Eyed Wisdom". The Theosophists and early Kabalists associated it with the bird of Saraswati, the goddess of Wisdom; Karttikeya the Kumâra, and the Virgin celibate of the Mysteries of Juno.

Helena Blavatsky defined it as "The gods and goddesses connected with the secret learning."

II. Etymology:

Arabic origin and pronounced Malak Ta-wooth.

Also called Malak Taus in Arabic and Turkish, and Tawûsê Melek in Kurdish.

Epistemologically, Muluk derived either from the Arabic Malaak (Angel), or Malak (King); same words in Hebrew. Muluk-Taoos is an ancient term used by the Yezidis, Persians, Arabs, Pakistani, Afghanistani, and several other ethnic groups in Anatolia, the Near East, and the Middle East.

Esoterists, cosmogenists, and theologians believe that Muluk derived from Malak or Maluk, "Ruler", a later form of Moloch, Melek, Malayak and Malachim, "messengers", angels.

III. In Ana'kh/Ulemite literature:
The meaning of Muluk-Taoos occupies a privileged place in the Anunnaki and Ulema texts, for it symbolized the balanced power. A power in perfect harmony with beauty.

- **a**-Muluk is Mulku or Malki in Ana'kh, and both words mean authority, sovereignty, and a strong command.
- **b**-Taoos as a word, is not mentioned in the Ana'kh. But later on in history, the Ulema who lived in the Middle East and the Near East borrowed it from the archaic Arabic word "Ta-wooth", which means a peacock, and associated the peacock with elegance and beauty.

Thus, the general meaning became the harmonious or beautiful strength or authority. For such authority or a strong command to become beautiful, the ruler or the king (Gal) must show his people justice, affection, and love for arts and science.

*** *** ***

89. Anunnaki's Plasma Viewers "Nin.nin.zar"

⌘ ⌘ ⌘

Definition:

The Anunnaki don't have films like the ones we have on earth. In that context, they are more advanced and their techniques are complex.

Instead of rolls of films in cans, DVDs and discs, the Anunnaki have "Plasma-Viewers"; they are huge screens that they move from one area to another with a click from their mind.

Those screens are some sort of magnetic plasma projections, and can be reproduced on any surface, including walls, tables, even non-physical, non dimensional substances and materials.

The themes of their "films" are of a documentary nature and include arts, nature, science, space, and even the future. They do have plays, like the Greeks. They love orchestras. The musicians are dancers too, and they participate in illustrative movements.

*** *** ***

90. Elevation of Anunnaki's knowledge "Nizrin"

⌘⌘⌘

Definition:

Nizrin is the process by which an Anunnaki elevates the contents of knowledge recently deposited in his/her mind through the "conduit".

The ascension phenomenon occurs frequently and periodically, so the Anunnaki individual could catalogue the categories of knowledge recently acquired.

Elevating the contents is a mental projection that begins from within the brain's cells and materializes on a memory pad that appears before an Anunnaki

*** *** ***

91. Purification of an Anunnaki Student
⌘⌘⌘

I. Definition and introduction:

All Anunnaki students are required to purify their bodies before their orientation or their regular course of studies in the Anunnaki Ma' had (School; academy). On Earth, it is similar to the Jewish scribes, since earliest times, who had to purify their body before adding the name of God into the Torah?

It is the same principle, but different from the Jewish Mikvah. Each student must purify his or her own being alone, because the mixing of the impurities might produce a barrier to the proper purification.

Incidentally, remember the Essenes, these Judaic sect members of the Second Temple era?

At first, they used the Anunnaki's style of purification, but as time went by, and their numbers grew, they changed into a Mikvah-like, communal purification. A certain similarity can be established with the Christian baptism; it is all the same idea of purity and cleanliness.

The Christians believe that the mind and spirit are indeed cleansed by the baptism.

II. The purification process:

Anunnaki-Ulema Sorenstein explained the process.

He said verbatim:

The purification process/exercise occurs inside a small room, entirely made of shimmering white marble.

49

1- The purification room:

- In the middle of the room, there is a basin, made out of the same material, and filled with a substance called Nou-Rah Shams.
- Nou-Rah Shams is an electro-plasma substance that appears like a 'liquid-light.' It actually means, in Anakh, The Liquid of Light.
- The substance Anunnaki use purifies the mind and spirit as well. It's all very pleasant.
- Every minute you spend in the basin, you will feel lighter, happier, more complete within yourself and sparkling clean.
- Nou, or Nour, and sometimes Menour, or Menou-Ra, means light. Nour in Arabic means light. The Ulema in Egypt, Syria, Iraq and Lebanon use the same word in their opening ceremony. Sometimes, the word Nour becomes Nar, which means fire.
- This is intentional, because the Ulema, like the Phoenicians, believed in fire as a symbolic procedure to purify the thoughts.
- Shams means sun.

2- Nif-Malka-Roo'h-Dosh":

- The most important phase of the purification is called Nif-Malka-Roo'h-Dosh".
- Nif means mind. Centuries later, it was used by terrestrials as Nifs or Roo'h, meaning Soul or Spirit. Since the Anunnaki did not believe in a 'separate soul,' the mind was the only source of creation and mental development, while humans continued to interpret it as 'Soul.' It means the same thing in Akkadian, Hittite, Aramaic, Hebrew and Arabic.
- Malka means kingdom or a higher level of knowledge and mental development. Humans changed Malka to Malakoot or Malkout; and the same, or very similar word, was again used in Aramaic, Hebrew, Syriac, Coptic, Arabic, Phoenician and so many other languages.
- Roo'h is the highest level of mental achievement.
- The Arabs use the same word, and the Hebrew used the word Ru'ach. However, the meaning changed in both languages, to represent soul, not mind.

- Dosh means revered.
- After this phase of purification, the Anunnaki student enters a moderately sized room off the classroom.
- In the room there is a cell, shaped as a cone and transparent.
- The cell floats in the air, approximately twelve centimeters above the floor.
- The top of the cell is connected to a beam originating from a grid attached to a ceiling floating in the air; it is totally suspended on its own.
- The students steps inside the cone.
- A door opens and the student enters the contraption.
- Inside the cone, a clear fog begins to form in the center.
- After a short while, the fog's color changes from white to silver-blue in form of waves, and the student starts to see his/her thoughts registering on a machine serving as an information board and which is posted on the right side of the cell.
- These thoughts begin to take physical shape, which is instantly copied to a screen.
- The screen transforms the thoughts-form into a code.
- Almost instantly, the code is transformed into a sequence of numerical values.
- The sequences are the genetic formula of the student.
- This genetic formula is the "Identity Registration" of the Anunnaki student. In terrestrial terms, you can call it DNA.
- But it is more than that.
- It is the level of mental readiness for the next stage.
- At this point, the student begins to hear a direction in his/her head, as clear as if someone is talking to him/her directly. This is the moment when the student has been approached on a telepathic level.
- The student is instructed to free his/her mind from all thoughts.
- It is something like what the Japanese call "Koan," or "Kara," a state of "mind nothingness."
- Then, something important starts to happen; rays of various densities and colors surround the student in a

51

cloud. It is tempting to compare it to the aura in terrestrial terms, but this is not the case.

- It is not an aura, because it is not bio-organic. It is entirely mental.
- What happens next takes only one minute; this is the most important procedure done for each Anunnaki student on the first day of his/her studies – the creation of the mental "Conduit."
- A new identity is created for each Anunnaki student by the development of a new pathway in his or her mind, connecting the student to the rest of the Anunnaki's psyche.
- Simultaneously, the cells check with the "other copy" of the mind and body of the Anunnaki student, to make sure that the "Double" and "Other Copy" of the Mind and body of the student are totally clean.
- During this phase, the Anunnaki student temporarily loses his or her memory, for a very short time. This is how the telepathic faculty is developed, or enhanced in everyone.
- It is necessary, since to serve the total community of the Anunnaki, the individual program inside each Anunnaki student is immediately shared with everybody.
- Incidentally, this is why there is such a big difference between extraterrestrial and human telepathy.
- On earth, no one ever succeeds in emptying the whole metal content from human cells like the Anunnaki are so adept in doing.
- Lacking the Conduit that is built for each Anunnaki, the human mind is not capable in communication with the extraterrestrials.
- However, don't think for a moment that there is any kind of invasion of privacy. The simplistic idea of any of your friends tapping into your private thoughts does not exist for the Anunnaki. Their telepathy is rather complicated.
- The Anunnaki have collective intelligence and individual intelligence. And this is directly connected to two things:
- The first is the access to the "Community Depot of Knowledge" that any Anunnaki can tap in and update and acquire additional knowledge.

- The second is an "individual Prevention Shield," also referred to as "Personal Privacy."
- This means that an Anunnaki can switch on and off their direct link, or perhaps better defined as a channel, to other Anunnaki.
- By establishing the "Screen" or "Filter", an Anunnaki can block others from either communication with him or her, or simply prevent others from reading any personal thought.
- "Filter" "Screen" and "Shield" are interchangeably used to describe the privacy protection.
- In addition, an Anunnaki can program telepathy and set it up on chosen channels, exactly as we turn on our radio set and select the station we wish to listen to.
- Telepathy has several frequency, channels and stations. When the establishment of the Conduit is complete, the student leaves the conic cell and heads toward the section assigned to him or her at the classroom.

*** *** ***

92. The Highest Class of the Anunnaki "Rafaat'h"
⌘ ⌘ ⌘

I. Definition
II. Anunnaki and the concept of God

I. Definition:
Pronounced rough-ah-ath. The highest class or category of the Anunnaki, which is ruled by Baalshalimroot.

This class is considered to be the elite, because of their very advanced knowledge of "Shama" (Universe) based on science and "Fira-Sa". The followers and subjects Baalshalimroot are called the "Shtaroout-Hxall Ain," meaning the inhabitants of the house of knowledge, (Mistakenly, the Aramaic and Hebrew texts refer to as: House of God) or 'those who see clearly."

II. Anunnaki, Phoenicians, and the concept of God:
At one point in ancient times, the Anunnaki Rafaat'h told the Phoenicians that there is no god (One God) ruling over the entire universe.

However, the high priest of Melkart (Chief god in Tyre, Carthage and many regions in the Near and Middle East) instructed the temple's priests to mislead the people, and spread the word that the Anunnaki were celestial gods visiting Earth, and are constantly working with the Phoenician gods and priests.

In the early tales about Kadmos (Kadmus), the Phoenician prince who lived around 2,000 B.C. according to Herodotus of Halicarnassus (482-B.C.-426 B.C.), the concept of one god instead of many gods began to surface.

It was based upon the belief that the Anunnaki followed one supreme leader who created the entire human race. But even then, the term "god" did not mean the "God" we worship today.

93. Satana-il and his Subjects
⌘⌘⌘

Also called Satan-na-il or Shatan-Il.
I. Definition and introduction
II. Angels of God

I. Definition and introduction:

Satan-na-il, also called Sa-tan, Shatan or Shaitan means the devil, the bad angel. Shaytan, in Aramaic, Phoenician, Syriac, Ashurian (Ashuri, Assyrian), and Arabic means the devil. In Pre-Islamic and Islamic literature as well, Shaytan is Lucifer, also called the Dijjal (The Impostor).

"Il" means god. Thus Satan-na-il (with all its derived names, and linguistic variations such as El, Eli, Al, etc.) becomes: The god of evil. But epistemologically, he is the god of the fallen angels. Early doctors of the Eastern Church confused him with Baal-Zebub.

More confusion will arise in the early literature of Eastern Christian Church, when a reference was made to a war waged by the angels of God (Judeo-Christian God) against the fallen angels of Satan-na-il.

Satana-il was the supreme leader of an extraterrestrial race that accompanied the Anunnaki in their second landing on earth. This galactic race was physically and genetically different from the Anunnaki and the Igigi.

Their duty was to serve the Anunnaki. They rebelled against the Anunnaki and broke the laws of their leader by breeding with the women of the Earth. Contrary to the general belief, the Anunnaki were not the first extraterrestrial race to marry, or the have sexual relations with the women of earth.

The subjects of Sata-na-il were the first to take the Earth's women as their wives. But because they were integrated into the

Anunnaki's community, many believed that they were Anunnaki themselves. According to the Sumerian mythology and the Bible, their sin and breaking the laws of the Anakh caused the Deluge. In many parts of the Book of Enoch, they were mentioned as the fallen angels.

II. Angels of God:

Those angels were:

- **1**-Gib-ra-il (Angel Gabriel), the guardian of Janat Adan or Edin (Garden of Eden), in Sumerian and in Anakh is Nin-il, or Nin-Lil. It is also called "Gab" and "Gab-r-il". Gab means a female guardian, a governor or a protector.
- **2**-Mi-Kha-il (Angel Michael), the Christian "Guardian angel", also known in the civilizations of the Sumerians, Babylonians, Acadians, Hittites and Anakh as Nin-Ur-ta.
- **3**-Rapha-Il (Angel Raphael), known to the Sumerians and Anunnaki as Enki or En Ki.
- **4**-The other angels of God were Raguel, Sariel, Ramiel, and Uriel, known to the Sumerians and Anunnaki as Enlil or En-Il.

Worth mentioning the fact that the archaic terms "Il", and "El" were understood sometimes as angels, prior to the writing of the Semites, Phoenicians, Ugaritic, Hittites and Acadians epics and mythologies. In the original Sumerian and Akkadians texts, "Il" or "Eli" or "Ili" meant =high, elevated.

This explains why Angel Gabriel was represented to us as the guardian of the Garden of Eden. In the ancient texts of the Sumerian, Akkadians and civilizations of the neighboring regions, Gab-r was the governor of "Janat Adan" (Garden of Eden). But "Angel Gabriel, the Sumerian is more than a guardian, because he was called Nin-Ti which means verbatim: Life-Woman.

In other words, Angel Gabriel was three things:

- **1**-Governor of the Garden of Eden;
- **2**-A woman, not a man, because she was "the female who created life";
- **3**-A geneticist who worked on the human DNA/creation of the human race.

94. Shamrakh-Ank-Sinhar-baal
⌘ ⌘ ⌘

I. Definition and etymology
II. In Eastern languages

I. Definition and ethymology:
Shamrakh-Ank-Sinhar-baal is the creator of life in the universe, and one of the primordial creators of the Anunnaki.
Derived from the Ana'kh (Anunnaki language).

It is composed of:
- **a-** Sham, which means light, fire, sun.
- **b-** Rakh, which means a field, an area, species.
- **c-** Ank, which means life, energy, creation.
- **d-** Sinhar, which means a leader, a lord, a commander.
- **e-** Baal, which means a creator, an inventor, a scientist, god.

II. In Eastern languages:
Later on in history, many Eastern civilizations will borrow some of these words and incorporate them in their native tongues and languages.

For instance (To name a few):
- **a-** The Egyptians took the Anunnaki's word "Ank";
- **b-** The Phoenicians, Habiru, Akkadians, Hittites, the scribes of Aramaic and Syriac, and Arabs took the Anunnaki's word "baal" and transform it into Baal, Bal, Al, El, Allah, Eli, Elohim, Ilah, Il, Eil, etc...

59

95. Tau "Taw"
⌘ ⌘ ⌘

Definition:

Anunnaki word/representation of the geometrical figure of the cross. Taw was incorporated in the Phoenician, Hebrew and Ulemite languages.

- The Anunnaki's Taw became:
-
- **a**- The square Hebrew letter "Tau";
- **b**- The Egyptian "Ankh";
- **c**- The Latin Crux Ansata;
- **d**- The Masonic symbol for the Royal Arch Degree;
- **e**- La Croix des Templiers (Knights Templar Cross);
- **f**- The Maltese Cross of the order of St. John;
- **g**- The Mexican astronomical cross found in palaces in Palenque;
- **h**- The Indian sacred mark placed on the brows of their Chelas.

*** *** ***

96. Ti

⌘ ⌘ ⌘

Definition:
An Anunnaki, Sumerian and Babylonian word for rib. In later versions of the ancient Akkadian/Sumerian texts, "Nin-ti" became the "Lady of the rib", also the "Lady of life", and the "Lady of creation".

Consequently, Adama (Adam), the man, was created from the rib of Ninti (Also identified with Gabriel, the female Anunnaki goddess/angel and life giver.)

This, contradicts the story of the creation of Adam and Eve as told in the Judeo-Christian tradition. According to a Babylonian myth and Ana'kh texts, a woman created man; it was not a man who created a woman (Eve).

And the female Anunnaki (Nin.Ti/Gabriel) used her rib to create Adam.

The early translators, (and possibly, intentionally misleading or creative scribes) of the ancient texts and epics of Sumer got it wrong, and their fake story of the creation of Adam and Eve invaded the Hebraic, Christians and Islamic holy scriptures.

*** *** ***

97. Children of Anak (Anakim) "Zamzummim" ⌘ ⌘ ⌘

I. Definition
II. In occult and esoterica
III. Benefits
IV. Geometrical presentation

I. Definition:
A name for the children of Anak (Anakim) as referred to by the Ammonites.
Zamzummim derived from the Phoenician Zayin. It has the same pronunciation in Aramaic, Syriac, Hebrew, and Arabic.
In Ana'kh it is Za. YIL, and means weapons.

II. In occult and esoterica:
Code/Use according to mythology and esoterism: To be written seven times on a small stone and placed inside the house, more precisely in the foyer of the house. On the road, it is advised to grab it in your right palm, and repeat the name of the opposition seven times.
Once done, you keep it in your right pocket.
During mental communication, you draw a circle and you place the stone on the left side of the circle, facing north.

III. Benefits:
*   **1**- Defense against intruders;
*   **2**- To overcome or change a negative decision by a third party that can affect your well-being and/or assets;
*   **3**- Helpful in negotiations, and while giving a speech, lecture or a presentation;
*   **4**- Eliminate negativity.

IV. Geometrical presentation and symbol: Two adjacent triangles.

65

98. Anunnaki's Physical Manifestation "Zouhoor"
⌘⌘⌘

I. Definition and introduction
II. Ambar Anati explains the phenomenon

I. Definition and introduction:

Anunnaki's physical manifestation is a rare occurrence.

And very few people witnessed it. Anunnaki's manifestation takes place always in the dark.

Particles of lights are usually seen around unclear formation that gradually grows to form a humanoid body-look-like. Then the body readjusts itself to adopt a human body shape. A small group of ufologists called "Mystique Ufologists" calls it "Divine Visitations", or "Heavenly Visitations".

It is obvious that some sort of religious influence plays part in that definition.

The reality is quite different.

II. Ambar Anati explains the phenomenon:

Sinhar Ambar Anati, known to us as Victoria, the hybrid wife of an Anunnaki who lived among us for a while, commented as follows:

- Alien's Manifestation has nothing to do with faith or religion.
- Zoohoor is simply the first stage of an extraterrestrial presence in an electro-plasma body gathering itself to adopt the shape of a human body.
- The Anunnaki are known to appear that way.

67

99. Hell
"Jahaan"
"Jahaam"
⌘ ⌘ ⌘

I. Definition
a- Etymology
II. Ulema's view
III. Jahaan "Hell" in the ancient Near East literature

I. Definition:

An Anunnaki/Ulemite term for an afterlife dimension/sphere, which could be the equivalent of the concept of Hell, called Jahannam in Arabic, Gehenna, Gehenom, and Gehinom in Hebrew, Gehenna and Jahenem in Aramaic, and Inferno in Dante's Divine Comedy. However, the Ulemite Jahaan is quite different from the general concept of Hell in all major religions.

a- Etymology:

It is obvious that the Hebrew words Gehenna, Gehenom, and Gehinom and the Aramaic words Gehenna and Jahenem for hell derived from the Anunnaki's word Jahaan. However, the word "Gehenna" is not used exactly the same way in the Holy Scriptures, even though all Hebrew and Aramaic words are translated as the words for "hell" in the King James version of the Bible.

In the time of Jesus Christ, the word Gehenna was used in reference to the Valley of Hinnom where trash was burned.

Jahaan and Gehenna became Jahannam in Arabic. They are the names of Hell.

According to the Ulema, there is no Hell and no Heaven, but a multitude of spheres of existences in the afterlife, that include humans, animals and various life-forms.

69

II. Ulema's view:

- The righteous people will be reunited with their loved ones including their pets in the afterlife.
- This reunion will take place in the ethereal Fourth dimension.
- The reunion is not of a physical nature, but mental.
- This means, that the mind of the deceased will project and recreate holographic images of people, animals and places.
- All projected holographic images are identical to the original ones, but they are multidimensional.
- Multidimensional means, that people, animals and physical objects are real in essence, in molecules, in DNA, and in origin, but not necessarily in physical properties. The physical substance of the body becomes an "image".
- In other words, what you see in the afterlife is real to the mind, but not to your physical senses, because in the after life (In all the seven levels/dimensions of life after death), physical objects, including humans' and animals' bodies acquire different substances, different molecular compositions, different properties, and newstructural forms.
- The physical rewards and punishments are mental, not physical in nature, but they are as real as the physical ones.
- The deceased will suffer through the mind.
- The pain sensations are real, but are produced by the mind, instead of a physical body. So in concept and essence, the Ulema and Hebraic scholars share similar beliefs; the good person will be rewarded, and the bad person will be punished.
- For the Jews, it is physical, while for the Ulema it is mental, but both reward and punishment are identical in their intensity and application.
- The wicked will not be indefinitely excluded from a reunion with loved ones.

- The wicked will remain in a state of loneliness, chaos, confusion and mental anguish for as long it takes to rehabilitate him/her.
- This state of punishment and rehabilitation can last for a very long period of time in an uncomfortable sphere of existence inhabited by images of frightening entities created by the mind as a form of punishment.
- The projection of these macabre and scary entities are created by the subconscience of the wicked person. Other scholars believe that the holographic imageries are produced by the Double housing the mind.
- Eventually, all wicked persons will reunite with their loved ones after a long period of purification and severe punishment.
- Soul is a metaphysical concept created by Man.
- Soul is a religious idea created by humans to explain and/or to believe in what they don't understand.
- It is more accurate to use the word Mind instead.
- The mind thinks and understands. The soul does not, perhaps it feels, if it is to be considered as a vital force and source of feelings in your physical body.
- In the afterlife, such source of feelings is non-existent, and in the dimensions of the after world, such source is useless.

III. Jahaan "Hell" in the ancient Near East literature:
The earliest literary evidence for the Ancient Near Eastern Jahaan (Underworld realm) is found on the clay tablets that preserve the region's written culture, such as:

- **a**-"The Dream of Enkidu";
- **b**-"The Huluppu Tree" episodes in the Sumerian Epic Gilgamesh (2000–1400 BCE);
- **c**-The Sumerian Descent of Inanna to the Netherworld (c.1900–1600 BCE);
- **d**-The Canaanite/Ugaritic story of Baal and the Underworld (1675–1545 BCE);
- **e**-The Akkadian Descent of Ishtar (c.1100 BCE);
- **f**- The Assyrian Vision of Kummâ (mid-seventh century BCE).

100. Afarit
⌘ ⌘ ⌘

I. Definition and introduction
II. Some of the most important names are
III. Names of Afarit
IV. The "Four Heads"

I. Definition and introduction:

Arabic. Noun. Derived from the Ana'kh Afa-rit, meaning the entities with extraordinary powers. Afarit are usually associated with talismans. According to the Book of Ramadosh, the Afarit were genetically created by the Anunnaki, and were referred to as the "Iblis Forms".

Tawfic Canaan stated, a Talisman is a small amulet or other object, often bearing magical symbols, worn for protection against evil spirits or the supernatural. Demons are ordered in talismans to follow the instructions and to leave the patient whom they inhabit. A spirit of the lower world is assigned to every day of the week.

II. Some of the most important names are:

- El Mudhib, known as abu 'Abdallah Sa'id rules over Sunday;
- Murrah el-Abiad abu el-Hareth (Abu n-Nur) over Monday;
- Abu Mihriz (or abu Ya'qub) El Ahmar rules over Tuesday;
- Barqan abu l-'Adja'yb rules over Wednesday;
- Shamhurish (el-Tayyar) rules over Thursday;
- Abu Hasan Zoba'Ah (el-Abiad) rules over Friday;
- Abu Nuh Meimun rules over Saturday.

III. Names of Afarit:

Some authors think that these names are only synonyms to those of the four archangels.

The names of the afarit are:

- **a**-Damriat (Tamriat) for Mudhib,
- **b**-Man'iq (or San'iq) for Meimun,
- **c**-Wahdelbadj (or Wahdeliadj) for Barqan,
- **d**-Soghal for el-Ahmar.

IV. The "Four Heads":

El-Ahmar is also called Abu Tawabi, the father of all tawab. (Plural of tabi, the masculin of tabi'ah is qarineh.) The names of the "Four Heads" (Al-arba ru'us), are also called the "Four Helpers" (Al-a'wan Al-arba' ah). They play a very important role in talismans.

They are:

- **1**-Mazar the lord of the East,
- **2**-Kamtam the lord of the West,
- **3**-Qasurah the lord of the South,
- **4**-Taykal the lord of the sea.

Many of the names discussed above show clearly Ana'kh and Hebrew influence. With the exception of a few rules there is no method whatsoever to help in determining the origin or the way of forming the names of the supernatural powers, without an intense study of the Kira'ats (Readings) and teachings of the Ulema.

*** *** ***

100. Link to the Beginning
⌘ ⌘ ⌘

Master Ghandar's Kira'at:
The Honorable Master, Ulema Ghandar, in Ulemite/Ana'kh, (Anunnaki language) said verbatim, word for word, and unedited, in a 1963 Kira'at (Reading/Lecture):

- Don't feel threatened or humiliated by the expression "as it has been decided upon by the Anunnaki."
- There is no reason whatsoever, to think that we are the slaves of an extraterrestrial race.
- Sometime, perhaps always, some major decisions by a more intelligent, advanced and responsible group or an elite are necessary, to keep order, peace, and justice in the universe, in our lives here on Earth, in our societies, and even in our relationships with others. Otherwise, our lives and the stability of our very existence will be seriously jeopardized, and chaos will disrupt all forms of progress, happiness, and success in our endeavors.
- Consequently, decisions must come from those who are superior to us, from those who are more intelligent than us, from those who really care about us, unconditionally, no strings attached.
- So far, humans were/are not able to do so.
- As long as, here on planet Earth, we have markets' competitions, different understanding of what is right and what is wrong, as long as we have so many different religions, each one claiming to be the most truthful religion, and as long as, using aggression (in any form or for any reason) to convince others, to intimidate others, to rule others, and to protect interests at a personal level or at a state level, as long as we are controlled by these rules and differences among us, the human race will

75

remain negatively affected, greedy, violent, selfish, materialistic, and unable to seek and reach happiness, peace and the ultimate truth.

- Only the Anunnaki are capable and entitled to maintain law, order, peace and stability for us, for our families, for our children, and for our future.
- The Anunnaki know best, because they have created all life-forms on Earth; from the smallest fish in the oceans to the biggest idea or concept a human being can think of.
- Materialistic things, including ideas, concepts, dreams, joys, fears, and aspirations, all of them were originally created and propagated in our minds, bodies, and actions, at the time the Anunnaki conceived a "picture", a structure, a substance, and a permanent physical-mental nature of the human race.
- They created us thousands of years ago, long before Man used the word "God", and followed any religion.
- Here, I am talking exclusively about the Anunnaki, and not about other extraterrestrial races, and they are so many all over the universe.
- When it comes to our place, role and future on the landscape of the universe, and especially our relation to the Creation, Creator or Creators, life after death, and our possible existences in other worlds, there is a big difference between the Anunnaki and other extraterrestrial races.
- Because the Anunnaki have shaped us physically and mentally, they looked upon us as their cherished creation.
- Here, there is a direct relationship between us and them.
- In fact, a part of us resembles a part of them. And we shall talk about this resemblance very soon.
- On the other hand, extraterrestrial races who had nothing to do with us at any level, and for any reason, saw us as an inferior species; inferior in so many ways.
- Thus, these "stranger extraterrestrials" care less about us, about our families, and the continuity of the human race.

- In the ancient past, these "stranger extraterrestrials" interacted very briefly with early forms, early quasi-humans, and early "categories" of human beings.
- Nothing great came out of it.
- They did not even bother to teach us anything.
- Would you teach an insect how to write a symphony, how to build a hospital, or how to plant a rose garden? No, you would not, because you believe that an insect is incapable of learning and creating such beautiful and intelligent deeds.
- Well, this is how exactly, extraterrestrials viewed our ancestors, and ironically continue to think about all of us.
- We are not worthy of anything, because on the galactic map, humans have so little to offer to the cosmos. This is not a negative attitude. This is not pessimism. This is a fact.
- But, the Anunnaki do not share at all the opinions of other extraterrestrial races.
- To them, we, mean a lot, but to a certain degree.
- But here, I must be totally honest with you, dearest friends and students.
- Look around you, everything you see has a past, a present, and a future.
- Look at all these beautiful plants in the fields. Their past is seeds. Their present is what you see. Their future is to fertilize the earth, or to become something else by changing into something new, different, another organic substance, so on.
- Well, this was not the case for humans, when humans first appear on Earth.
- There are so many opinions and ideas about our origin on Earth. And we will not succumb to rhetoric.
- Some of the early human races vanished for many reasons.
- Others survived and relatively developed into a more intelligence species.
- Others became us, no matter what science calls us.
- The human category that has survived and became us, is essentially a sort of a genetic animal-human blend.

- Yes, we were, and still are a genetic formula.
- And as you know, a formula does not need a past to exist, only an idea at the beginning, followed by research and experiments, leading to the creation of something new.
- This is exactly what happened to us; how and why the early forms of humans came to exist, and develop into something else, without a definite and a well-prescribed future. And I will explain this, because it has a major impact on our future.
- Before I continue with the Dirasa, I want to explain to you, briefly, something very important. And keep it in your mind.
- Everything you see in the universe came from something else.
- Not necessarily from the same material or the same substance.
- Nothing in the Donia (World) appears for no reason.
- Nothing in the Donia, comes from nothing, except the beginning of the universe itself. And I will talk about this, later.
- Always, and always, keep in mind, that there is an origin for everything.
- And this origin, or as we call it "Source" is also what we call in our study "Khalek" (Creator).
- Creator is in fact "Creators".
- In the weak mind of humans, what appeared to be (To our ancestors, prophets, and creators of our religions on Earth) the real, unique, and only God, was in fact, the "strongest" and most powerful god from all the other gods (Humans came to know, or heard about) who managed to outlaw other gods, to eliminate other gods, and convince our ancestors, that He is above all gods; the only one; and the creator of human beings, and the universe.
- Struck with fear, and limited in knowledge, science, and history of the universe, our ancestors (From all the nations, and different religious beliefs) accepted that "god" as the real god of the universe, and creator of the human race.
- The Habiru (Israelites; Hebrews) called him Yahweh.

- The Phoenicians called him El; Baal.
- The early Christian Arabs and Muslims called him Allah.
- The English speaking nations called him God.
- The French called him Dieu, so on...
- The truth is, the "Origin", the "Source", and the "Creator" of humans, and life as we know it today, are in fact one and the same: Creators.
- The word "Creators" means, those who generated life on Earth.
- In other dimensions, they are called differently, but we are not going to worry about that, for now.
- The "Creators" were a very advanced extraterrestrial elite who used to travel many parts of the universe. A more accurate word would be the "Multiverse". And Earth is one "layer" or one single dimension of the "Multiverse".
- Thus, it is extremely important to believe in a "Creator", whether in the singular or the plural form.
- You look at your wrist-watch, and you instantly realize, that there is a watch-maker.
- You look at the car you drive, and you understand that there is an automobile factory.
- Thus, when you look at your body, you should understand, that there is a body-maker.
- Very good. Then, let's start with this body of yours, you are now looking at.
- This body of yours is made from cosmic dust.
- Everything you see on Earth, and around Earth is made from a cosmic dust.
- You can call this cosmic dust whatever you want; Cosmic rays, gases, molecules, particles, carbon, proton, neutron, atoms, bubble, so on.
- The cosmic dust was produced from the collision of several layers or space bubbles at the beginning of time and space.
- Part of this cosmic dust fell on Earth.
- The cosmic dust on Earth produced all sorts of things; mountains, rocks, strange and archaic life-forms, plants, clouds, and yes, the primordial and early forms of creatures that moved or walked on fifty legs, four legs, three legs, and without legs.

79

- Some were animal-plants; some were animals; some were something else; and others animals and animals-humans.
- All of them were produced and came from the cosmic dust of the exploded universe.
- At that moment in history, time and space, and beyond, the "Creator", the "Creators", and the God we know had nothing to do with humans, early humans, and animals-humans life forms.
- In other words, the human race (From its early archaic form to the present) was created by non "Creators", "Creator", "Judeo-Christian-Muslim God".
- And this applies also to all the extraterrestrial races and their habitats.
- The extraterrestrials like us, were also created by and from a cosmic dust.
- But they are different from us, physically, emotionally, psychologically, and mentally. And we will talk about this, some other time.
- So, when Earth was created, and during the very early existence, and following stages of Earth, some forms of animals and humans were created too.
- The early humans had a multitude of categories.
- Each category was shaped differently.
- Every shape depended on its origin/place of creation.
- For instance, humans or quasi-humans who came from the sea, looked very different from their counterparts who came from land.
- Some humans or quasi-humans were reptilians. At least quasi-humans, with the face of an archaic undeveloped human, with a long tail.
- Some walked on three legs.
- Some walked on four legs, and so on.
- No intelligence faculty was yet installed in those bodies.
- Call them robots or moving creatures if you want.
- They were a blending and a matching part of the ecology and landscape of Earth.
- And since the "Conduit", or an "intelligence faculty" was not an original part of their organic or bio-organic substance (Structural composition), a future or a reason

for reproduction was not necessary, even though, in many instances, materials, substances, and species can reproduce without intelligence.

- But this is exactly what makes the primordial and definite difference/distinction between humans and other living-forms (Life-forms).
- At that time in history, 37 different quasi-humans (half animals-half humans) lived on Earth. Some esteemed Ulema suggest 47 different species. We will talk about that in our next Dirasa.
- On other planets, in different galaxies, in distant dimensions, life evolved differently.
- Different kinds of life-forms were created.
- Some life-forms were the extraterrestrials.
- And some of these extraterrestrials were or became highly advanced. Not all of them.
- Some extraterrestrials who lived on planets similar to Earth, looked like modern humans, but not totally.
- Some were giants, others were short, and other groups were horrible-looking creatures.
- The only three races who shared some similarities with us were the Lyrans, the Anunnaki, and the Igigi, who were mentioned in the Book of Ramadosh, and the Babylonian clay tablets.
- Between 450,000 and 460,000 BCE, an immense Anunnaki's mother-ship and numerous spaceships hovered over the Earth.
- First, they flew over Madagascar, Brazil, Australia, and Central Africa.
- Then, they began to scan the lands in the Near East.
- An immense mother-ship will stay still in the air, some thousands feet far from the surface of the Earth.
- And a great number of small flying machines would exit from under the belly of the mother-ship, and head toward Earth.
- They were piloted by Anunnaki.
- Their missions were to probe Earth, scan the lands and waters, and locate certain minerals, natural resources, and some rich aquatic substances found only in the seas and the oceans of the Earth.

- The Anunnaki accompanied by their allies, the Igigi landed in Phinikya (Phoenicia; Modern Lebanon).
- The huge mother-ship remained in the sky and began to orbit Earth.
- The small spaceships landed on Earth.
- The mother-ship also functioned as an intermediary space-station between Earth, the Moon, and Mars.
- Bear in mind, that before visiting Earth, the Anunnaki have already established colonies on the Moon, and on Mars.
- Some traces of those colonies, and minor evidence(s) of the pre-historic Anunnaki's civilization, are still visible on the Moon and on Mars.
- As soon as the Anunnaki and the Igigi landed in Phinikya, they began to colonize the area, and build settlements.
- The settlements consisted of movable (Mobile) prefab living units and quarters. You can call them if you want prefabricated homes.
- The chief of the Anunnaki who was in charge of the colonies in Phinikya was called Aa-kim-Lu.
- His other names were Anoon Elah-Im; Anu. Il-Ohiim; Anu-Ela-Kim.
- Some of the earliest great findings of the Anunnaki and the Igigi were the Maha'rit; The Ourjouwan, and the Zaa'-faran.
- The Anunnaki and the Igigi spent many years in Phinikya; they built huge cities, housing facilities, and laboratories.
- Then, they found out that Earth was extremely rich with natural resources, and quickly realized that they needed a larger team of workers to mine and extract these riches.
- Years ago, while exploring other regions on Earth, the Anunnaki discovered heavily built creatures in Madagascar, Brazil, Australia, and Central Africa.
- Some of these creatures were beasts, some half humans-half animals, and others were reptilians.
- Here, I have to remind you that the Anunnaki did not create these creatures. They were here on Earth, long

- time before the Anunnaki and the Igigi descended on Earth.
- So, the Anunnaki went back to Madagascar, Brazil, Australia, and Central Africa, looking for these creatures as potential workers.
- You already know what happened next.
- They captured those horrible looking creatures, and tried to domesticate them, exactly as humans did, when they captured wild horses and dogs.
- So, they caught them in masses, and brought them to Saydoon, Tyrahk, Kadmosh, Adonakh, Ilayshlim, and Markadash, their colonies in the Near East.
- Those archaic creatures were not intelligent at all. Consequently, they were unable of fulfilling any task that required intelligence, or even a minimal understanding of what they were supposed to do.
- The Igigi decided to create (Manufacture) creatures capable of carrying and executing hard labor and continuous physical work.
- Well, what they created was not what they have hoped for.
- In fact, their creatures looked more horrible than their predecessors, and acted and moved like deformed beasts.
- In short, the Igigi's creation experiments were a total failure.
- Then, the Anunnaki interfered, and after many attempts, trials and errors, and new genetic experiments, they succeeded in creating a satisfactory breed and/or species of humans.
- To accomplish this, they had to slaughter an Igigi, take his blood, mix it with earthly elements, blend the whole thing with wet Tourab (Clay), and breathe into the mold.
- This genetic procedure was one of several, and varied genetic experiments and methods conducted and used by multiple Anunnaki gods and goddesses.
- The Book of Ramadosh and the Akkadian clay tablets tell us about other means processes, and elements, the Anunnaki used to create humans.
- Some experiments were conducted in Chimiti (Tube; containers).

- Some experiments required the impregnation of Anunnaki women (Goddesses).
- Some experiments required the semen of Anunnaki men (Gods).
- Some experiments required upgrading an existing quasi-human race. So on.
- The Book and the clay tablets even give us the names of the Anunnaki men and women who created our ancestors.
- But up to that moment in history, the creatures manufactured by the Anunnaki remained at a robotic level.
- Years later, the Anunnaki upgraded their creatures by adding 13 faculties in their brain.
- This addition allowed the early humans to communicate with their creators.
- These faculties are called Fik'r-ra-ma (A bulk of understanding and reacting.)
- Years laters, the Anunnaki added some (Unknown to us) aquatic bacteria such as fungi, much needed for the development of certain organs of the early humans.
- The Anunnaki kept on exploiting Earth, building more colonies, and adding new cities neighboring Saydoon, Tyrahk, Kadmosh, Adonakh, Ilayshlim, and Markadash.
- Among the new major cities were Baalbeck, Anfa, Sippar, Nippur, to name a few.
- The Anunnaki and the Igigi stayed on Earth for hundreds of thousands years.
- And then, for reasons we do not totally understand, they abandoned their operations and cities on Earth.
- At least, in Phoenicia, Babylon/Mesopotamia and surrounding areas.
- Then, years later, the Anunnaki and Igigi retuned to Earth.
- But this time, the Anunnaki decided to create a more advanced human race.
- And they did.
- They called that race Bashar.
- This happened around 300,000 BCE.

- Around 125,000-100,000 BCE, The Anunnaki created the early human women, and were called "Women of the Light"; they were the early female-forms on Earth.
- Thousands of years later, the people who lived in the Arab Peninsula and the lands bordering Persia, the United Arab Emirates, and India, began to call these women "The Women of Light", and those who were allowed to "mix with them" were called "The Sons or Children of Light".
- From this early human race, all humans came to life. The Judeo-Christian God had nothing to do with the creation of the human race. In other words, the God we know, revere, and fear today did not create us. Even the word or term "God" did not exist in the early stages of the existence of the human race on Earth.
- Around 65,000 BCE, the Anunnaki created the final form (Physically and mentally) of humans.
- This final form did not change much throughout the ages. Basically, it remains the same to the present day.
- For reasons we know, and reasons we don't know for sure, the Anunnaki got mad at the human race, and the supreme god/lord of the Anunnaki decided to wipe out the human race, from the face of the Earth.
- The Anunnaki mythology tells us that the Anunnaki created avalanches of Tsunami (Great Flood) that changed and altered the geography and locations of lands and mountains on Earth and killed almost everybody.
- Some humans were saved, because a group of Anunnaki elite interfered on behalf of our ancestors, and warned them of the forthcoming annihilation plan of the supreme god/lord of the Anunnaki.
- Earth was repopulated, and new cities were erected in many regions of the Near East and Middle East.
- This included the lands of the Persian Gulf, Hadarmoot, Bahrain, Qatar, Kuwait, the Arab Peninsula, Turkey-Anatolia, Sumer-Mesopotamia-Babylon, Phoenicia, the Fertile Crescent, Syria, Jordan and Palestine, etc...
- From 65,000 BCE to 7,000 BCE, many things happened on Earth.

- But what really interests us here, is the emergence of a new human race.
- This emergence occurred around 5,000 to 6,000 BCE.
- At that time in history, and for the first time in the annals of humanity, a human race with a developed mind, knowledge of arts and science, a capability of recording historical events, with tools to erect cities and develop agriculture, apparently surfaced from nowhere, out of the blue.
- No anthropologist or a theologian seems capable to explain the missing link between this new human race, and the anterior ones, and the sudden development of civilizations on Earth.
- Pyramids were built. Extraordinary cities like Baalbeck were erected. Immense plantations and agricultural farms flourished everywhere. Several coherent languages were heard. Laws were formulated, to govern societies, city-states, and far distant nations. Organized religions, pantheons, deities, worship rituals , and priesthood cast appeared everywhere. And no historian or an anthropologist succeeded in explaining the sudden appearance, and origin of humanity's culture, languages, and civilization.
- It was at that time in history, that finally the Anunnaki got their act together, and decided to create a new human race, capable of functioning, reasoning, and understanding what is right and what is wrong.
- It was at that very specific time interval, that the first day of humanity's history was recorded. Because, that was the moment, when the Anunnaki created us in their image, using their own DNA.
- This explains when, why, and how, the quasi-humans became overnight intelligent humans, and all of a sudden, cities, art, science, music, poetry and human drama came to life.
- It was neither evolution, nor that God who asked Abraham to sacrifice his son, who created Man. But, the Anunnaki's genetic creation of Man.
- If it was God, then why it took God so long to create us? To create a reasonably looking human race? Science, archeology and anthropology have proved to us, that

mankind was not created five thousand years ago. And the skeletons and remains of humans who lived millions of years ago were not in the image of the God of Abraham, Jesus and Mohammad. They were half-animals.

- And if it was evolution, then how can we explain the sudden "appearance" of languages and civilizations without a direct link from 7,000 BCE to the early days of humans on Earth?
- Why humans who lived before 7,000 BCE could not talk normally? Could not build Pyramids? Could not write? Could not understand human anatomy? And all of a sudden, without a gradual and systematic development, and a phase-by-phase progress, they start to talk all kinds of languages, build huge cities, sail the oceans, and perfectly understand anatomy and astronomy?
- Why?
- Because of the immediate and sudden intervention of the Anunnaki. They have jumped and catapulted our development in no time.
- Neither God, nor evolution had to do anything with the creation of the modern human race.

*** *** ***

101. Behemoth: The Greys
⌘ ⌘ ⌘

I. Definition and introduction
II. Behemoth, Behemah, and Baha'em in Anunnaki's texts

I. Definition and introduction:

Ulema Lafayette said that Behemoth is the Anunnaki term for the alien race, "The Greys".

It is also a Hebrew word. The word was mentioned in ancient Egyptian, Phoenician, Greek, Hebrew texts, as well as in the Anunnaki' book "Book of Ramadosh".

In ancient texts of the Near and Middle east, Behemoth is the a spirit of the desert, possibly derived from the Egyptian deity, Taueret, about whom the Greek historian, Herodotus, wrote. The term "Behemoth" in the Hebrew is the plural form of the very common "Behemah" referring to a beast of use to humans or a dumb animal. It is being used here, however, as a single entity. Behemah is an Anunnaki word meaning beings from a "lower dimension."

The expression lower dimension refers to an extra-galactic habitat of aliens of a destructive nature, recently known to us as abductors of humans. From the Anunnaki's word Behemah, derived the archaic pre-Islamic word "Baha'em", which means humans, more precisely animal-human type race.

II. Behemoth, Behemah, and Baha'em in Anunnaki's texts:

Behemah, and Baha'em in Anunnaki's texts: The honorable Anunnaki-Ulema Ghandar, one of the custodian of the "Book of Ramadosh", said verbatim, word for word, and unedited:

- "At the beginning, the Baha'em lived on Earth as a bestial race. They were half humans and half animals.

89

- The Anaki (Anunnaki) came to Earth and upgraded the Baha'em, by mixing their genes with Anunnaki's DNA.
- The Baha'em lived in Australia, Brazil, Madagascar, and Central Africa.
- The Anunnaki caught many of them and began to conduct genetic experiments on the Baha'em.
- The Baha'em were brought to Tyre, Sidon, Baalbeck, Anfeh and Byblos in Phoenician (Modern day Lebanon), to Arwad (Island of Arwad, a Syrian territory), to Eridu and surrounding areas in Sumer (Ancient Iraq; Babylon; Mesopotamia), and to the Arab Peninsula.
- The genetic stock, the Anunnaki created in the Arab Peninsula was called Baha'em. This stock which looked quasi humans, were in fact horrible looking creatures with an enormous physical strength.
- Baha'em had deformed bodies; they were beasts and had reptilian characteristics.
- They lived in the Arab Peninsula, but some moved to Dilum (Ancient name of Bahrain, sometimes referred to as the original Garden of Eden) and Ourdon (Arabic name of Jordan).
- Later on, as legend has it, the early Arabs and particularly the Bedouins called them the "Spirits of the Desert", and the first creation of the gods.
- Some of our brothers have suggested that this quasi-reptilian grayish looking race inhabited regions in the hollow Earth and underwater in the Pacific Ocean, and the Mediterranean Sea.
- From the waters of the oceans, they extracted their energy.
- A new race/species of Baha'em was re-engineered by the Anunnaki. And this new genetic race produced a new breed of humans who were both good and evil.
- This new race is the lost race of early humans, and genetic link to the early creation of mankind."

*** *** ***

90

102. Chaiturabi
(Lively clay)
⌘ ⌘ ⌘

The Anunnaki's creation of Man from clay

I. Definition:
Composed of two words:
a-Chai, which means life.
b-Turabi, which means dirt; dust; earth; soil; clay.

From Turabi, derived the ancient Arabic word Turab, which means exactly the same thing. From the Ana'kh Chaii, derived the Hebrew word Chay or Chai (Pronounced like Jose in Spanish), which means life; when we add L', the meaning becomes "To life".
The Hebrew/Jewish "L" is similar to the Arabic "L", which means "to", in both languages, and sometimes, it means "toward" in Arabic.
In the first edition of the Aramaic/Arabic Bible, we found this sentence: "Anta min al turab, wa ila turab ta'hoodou."
Translated verbatim: "You are *(anta)* from *(min)* dirt *(turab)*, and *(wa)* to *(ila)* dirt *(turab)* you shall return *(ta'houdou)*.
The general meaning of Chaiturabi is the genetic creation of Adama species (First humans; primitive beings; human prototypes...) by the Anunnaki, and in-vitro fertilization by Ninti, Enqi and other goddesses in a laboratory called "Chimti". This creation was mentioned in Sumerian, Akkadian, and Assyrian texts.
The idea or concept of the creation of mankind from clay appeared in many mythologies, cosmogonies, and Ugaritic, Phoenician, Babylonian, Chaldean, Mesopotamian. Assyrian, Sumerian, Ana'kh, Ulemite, Greek, African, Aztec, Mongolian, Egyptian, Islamic, and Christian religious texts.

91

103. Exodus: The Hebrew Story Versus the Ulema Version
⌘⌘⌘

I. Definition and introduction
II. The controversial and colorful historical version/approach of a minor Ulema
III. For genetic reasons, the Anunnaki wanted a group of people
IV. The Levite Tribe
V. For the Anunnaki, women are the only ones associated with procreation
VI. An Anunnaki contacted Miriam
VII. The Immaculate Conception pattern
VIII. Moses was hardly even aware of the expedition
IX. The Phoenicians had the same genetic pattern as the Israelites
X. Yahweh, and the Anunnaki's spaceship
XI. Jewish women and Anunnaki hybrids

103. Exodus: The Hebrew Story Versus the Ulema Version
⌘⌘⌘

I. Definition and introduction:

From a joint writing project with Dr. I. Arbel: Exodus is a Greek word, which means a departure or emigration in a large number, or in groups. It is composed of two words:

a-Exo, meaning out; outside, and
b-Do from Dromo, meaning road; street.

Usually, it is referred to and it is associated with the Hebrew "Habiru" Exodus "Exit" from Egypt (circa 1500 B.C.) mentioned in the Bible.
The Book of Exodus depicts God Yahweh's call to the people of Israel to leave Egypt, and to head toward the Promised Land. The major players or key figures in the Hebrew Exodus are Moses, Aaron, Miriam, Pharaoh, Pharaoh's daughter, Jethro, Joshua.

II. The colorful historical version/approach of a minor Ulema:

To begin with, Moses never left Egypt. The Exodus as we know it, the flight of a large group of slaves who were saved by Moses and Yahweh from the evil Pharaoh, never happened. The Israelites were never slaves. And Moses, who was an Egyptian-Anunnaki hybrid, enjoyed an illustrious career as a general in Pharaoh's army.

He married a noble lady from a Hyskos origin, not the daughter of Jethro from the desert, and had several children, the boys following in his footsteps in the military, the girls marry noblemen. Who, then, is Aaron?

Where is Moses' sister, Miriam, who watched over him as he floated in his reed basket into the arms of the daughter and sister of Pharaohs, who decided to adopt the little infant?
What happened to his tribe, the Levites?
An Ulema adept said: "We are so used to the Bible that we accept it as either history or religion, but we generally don't analyze it. Think for a moment from a different angle.
First, the story of a child floating in a basket, and then saved by an important personage, is so common to many myths and religions, that it is almost a cliché.
Second, would the daughter of Pharaoh adopt a child with no name, possibly the offspring of slaves? Perhaps, you would say, she never saved anyone. She may have had an affair and gave birth secretly?
Honestly, would the proudest and most noble woman of Egypt, who could marry just about any prince on the face of the earth, demean herself like that?"

III. For genetic reasons, the Anunnaki wanted a group of people:

For genetic reasons, the Anunnaki wanted a group of people, whom we call the Israelites and who were living in Egypt, to come to Canaan at a certain date, and blend with the population. But the Israelites were not going anywhere. Why should they?
They were perfectly comfortable in Egypt, a sophisticated culture that not only allowed, but encouraged immigration and the blending of cultures. Some of them engaged in trades, others in agriculture, some did art and architecture, and many worked on

the building (Such as repairing and maintaining some areas of the pyramids, not the construction of the pyramids) of the pyramids for the Pharaoh in various capacities.

Incidentally, for the building programs, the Pharaoh did not need or want slaves. In an agricultural society, there are periods when the farmers have no work to do in the fields. During these times, many of them were more than happy to be gainfully employed by the government on the construction projects.

IV. The Levite Tribe:

The Israelites consisted of twelve tribes, and one of the tribes, the Levites, were the intellectuals, the philosophers, the teachers, scholars, etc. They were highly valued by the Egyptians.

The Levite tribe had connections with the Anunnaki before, so being contacted by them was never quite a surprise.

A young Levite woman, whose name was Miriam, was engaged by Batya, the sister of young Pharaoh, as a nurse/governess to her infant son, Moses. Third, where else, other than the Bible, does it show that the Israelites were *slaves* in Egypt?

Nowhere at all. More than that, there is not a single proof of the Exodus and of the invasion of Canaan.

So what really happened?

Another Ulema said: "Since I had been taught to emulate the Anunnaki in their ability to transcend time and space, and had studied these matters at first hand, let me tell you the story of the Exodus as it really happened, and set the record straight.

The controlling agent here is the Anunnaki, but you also must remember that the Exodus is allegedly headed, in the Bible, by a triumvirate:

- 1-Moses;
- 2-His brother Aaron;
- 3-Hisis sister Miriam. These two facts are the key.

V. For the Anunnaki, women are the only ones associated with procreation:

Moses was an exceptional child, and no wonder, since Batya had him with Anunnaki's help."

Historically, this has happened often in the royal Egyptian house; the Egyptians thought of the Anunnaki as gods, and the daughter of Pharaoh would consider it an honor to give birth to a child by the method that was removed from normal sex and considered to be an immaculate conception, exactly like the situation that happened thousands of years later:

- **a**-With Mary of Nazareth, who had given birth to Jesus, another Anunnaki hybrid;
- **b**-And Elizabeth (60 years old when she conceived) who gave birth to John the Baptist.

In Egypt, it was not even a secret, but something to be proud of. Miriam stayed in the palace with her young charge, teaching him, playing with him, and allowing the mother to attend to her State duties. She stayed with young Moses until he went to military school, at age seven, and then returned to her parents' home.

She was naturally expecting to find another excellent position rather quickly, since her qualifications, as nurse/governess to the Pharaoh's nephew were the highest in the land, but that was not her destiny. Now, you must adjust your mind to a totally non-Biblical fact. The Bible is male oriented, written and edited for a patriarchal society. But that was not the case with the Anunnaki.

For the Anunnaki, women are the only ones associated with procreation, and treated not only as equal, but as superior to the males. So what happened at that point might surprise the reader, if he or she is Jewish, Christian, or Muslim.

VI. An Anunnaki contacted Miriam:
An Anunnaki contacted Miriam, not Moses, and told her to go to the desert, and to visit Mount Sinai. This mountain was not very far, and a most convenient place for the Anunnaki to land their spaceships where it would neither disturb the people nor hurt their fields, buildings, or cattle.
The mountain had a flat plateau on top, surrounded by cliffs, a natural spaceport.
Miriam, who had known of many Anunnaki contacts, immediately set on her way to Mount Sinai.

On the mountain, she was asked to enter the spaceship, and was informed by the Anunnaki's commander who was her contact that her services were requested for the well-known pattern of Immaculate Conception, namely, impregnation by the Anunnaki system. Having accomplished that, she received, for safekeeping, a pair of small stone tablets, on which Ten Commandments were carved, a synopsis of a much larger set of laws that was carved on one of the most important Egyptian temples.

She was to hold on to these until further notice. Her mission, which was to take place in a few years, was to persuade a group of Israelites to leave Egypt and move to Canaan. While most would be reluctant, the Anunnaki hoped that some adventurous souls would be tempted by the opportunities offered in the new country.

Not that it was entirely unknown to them – their ancestors came to Egypt from this place, headed by the patriarch Jacob, whose son Joseph was a prominent official to a former Pharaoh. Miriam's hybrid child was a son, whom she named Aaron.

As expected, he developed supernatural qualities, common to Anunnaki hybrids. For example, he could make tricks with pieces of dry wood, and make them bloom with pink flowers, and he had the ability to turn wooden sticks into snakes, among other "magical" abilities.

When he was about ten, the Anunnaki contacted Miriam and told her the time had come to set out to Canaan. Miriam was by then married to a man named Jethro.

VII. The Immaculate Conception pattern:

As it always happens with Immaculate Conception personages, a typical husband had to be chosen. Somewhat older, kind and understanding, and perfectly willing to assist his wife and her hybrid son.

Joseph of Nazareth, the man who married Jesus' mother Mary, was exactly the same type.

To Miriam's surprise, the response to her suggestion of immigration was not bad. Mostly she was followed by Levites, but members of other Israelite tribes, less educated than the Levites, were greatly impressed by young Aaron's "miracles" and liked the idea of the riches they could find in Canaan.

The stone tablets, carved by the "gods," as they considered the Anunnaki, were also a great asset. And so a nice-sized group was created, not the thousands described in the Bible, but a few hundreds, and they set out to the desert without much delay.

VIII. Moses was hardly even aware of the expedition:
The Pharaoh never objected, there were no plagues or any other troubles, and Moses was hardly even aware of the expedition. Come to think of it, most of his time was spent out of Egypt, fighting for his Pharaoh with great distinction. As for Yahweh, he was not part of the picture at all. The Israelites, at the time, worshipped a whole pantheon, including:

- **a**-The Great Lion of Judah;
- **b**-Nechushtan the Snake, a shape-changing god who had enormous antlers and could become a stag at will; And the great Ashera, who had many names.

Yahweh was only one of them, and was mostly worshipped by the Levites, who seemed to enjoy his ferocity. They did not go directly to Canaan, but had to wait a few years, which they spent in the desert, acquiring the name "Habiru" which meant "Those Who Cross."
The reason for the delay was the need to synchronize their invasion with that of a young Phoenician commander, Joshua, Son of Nun.
Joshua, who had been described in the Bible as Moses' assistant, never heard of Moses in his life. He was privately connected by the Anunnaki, who wanted the immigration to take place simultaneously from two sides, the north and the south, in case the inhabitants of Canaan would object.

IX. The Phoenicians had the same genetic pattern as the Israelites:
Incidentally, the Phoenicians had the same genetic pattern as the Israelites, since in the distant past they lived in the same area and were related. But it so happened that the inhabitants of Canaan not only did not object to the immigration, they blended and mixed with the Israelites and the Phoenicians through many marriages, creating a genetic blend that was highly approved by the Anunnaki and is still strongly evident among today's Jews.

100

Naturally, when the "editors" of the Bible learned about this story, as they prepared to engage in the actual writing that was hundreds of years later.

They had to make changes, since allowing a woman to lead the Exodus would not do for a patriarchal society. So they switched the leadership to Moses, whose name was very well known even at that time, and made Miriam and Aaron his siblings. They made Jethro into the father of Moses' mythical wife, and then did their best to add a touch of drama and heroism – with great success.

One must admit that the Bible stories are roaring good tales about plagues, murders, fires, earthquakes, sacrifices, Deluge, rivers of blood, fury and vengeance of God, and cities' walls crumbling at the sounds of trumpets.

Everyone enjoyed these stories.

X. Yahweh, and the Anunnaki's spaceship:

Yahweh, who by then was the official God of the Jews, became the guiding light, and the Anunnaki's spaceship on Mount Sinai received a background of fire, thunder, and lighting, which beautifully fitted Yahweh's violent nature, and as for the burning bush and the other miracles, well, one must admit they make the story a whole lot more interesting.

XI. Jewish women and Anunnaki hybrids:

Of course, one of the most fascinating points of this history is that two women give birth to Anunnaki hybrids, using the scientific breeding system, and calling it an immaculate conception. And while Batya did not need to marry, Miriam did, and her husband, as we said, was exactly the type of husband Mary of Nazareth married. But are they the only ones?

Of course not. The Bible is full of them.

Think of Hannah. She is barren. She goes with her kind, older husband to the temple, and prays for a son.

In the Bible, a priest named Eli comes to her and promises her a son, if she will be willing to dedicate the boy to God. This does not hurt the boy in any way, on the contrary, little Samuel gets a very good education out of the bargain, but of course the priest was never a priest.

He was an Anunnaki, and he helped the barren woman get pregnant in his scientific way.

And what is the destiny of this child?

He is the one who would eventually anoint young David and make him replace the clinically depressed, homicidal, suicidal King Saul...And who is David, then? Young David is the son of Jesse, son of Obed, son of Boaz. And Boaz is the husband of Ruth. Yes, the famous Moabite lady who lost her husband to a plague and moved away from Moab with her beloved mother-in-law, Naomi, to the Land Judah.

What the Bible does not tell us is that the beautiful young widow, whose genetic makeup was just right, was contacted by an Anunnaki on the way to the Land of Judah, and with Naomi's consent, accepted the Immaculate Conception very gratefully.

Then, when settled in the Land of Israel and again with Naomi's consent, she meets Boaz, a distant cousin, a kindly, older man of the Joseph and Jethro type, who wants to marry her and give a good home to the two women and their little Anunnaki hybrid boy. Incidentally, Ruth lived to an advanced old age. So old, that she knew her great-grandson, David, and could smile at his incredibly good looks, complete with red hair, blue eyes, and glowing olive skin, which he had inherited from her Moabite side.

For once, her genetics won over the usually occurring black eyes and hair of the Anunnaki, and we know what a lady-killer David turned out to be. And they are not the only ones.

In the Bible, so many of the important female characters are barren and unable to conceive – until a very special person comes, sometimes as an angel, to help.

Sarah conceives at a ridiculously old age.

Rebekah needs her husband's Isaac special prayer to conceive – and probably a lot more.

Nothing can be more intriguing than the story of Elizabeth (Elisheva in Hebrew, meaning My God is an oath) who was the cousin of Mary of Nazareth.

There are various versions about her life, but amazingly, much of these stories are absolutely true. Yes, she was the cousin of Mary, daughter of a sister of Mary's mother.

She was married to Zecharaia, a good man who was typical to the Anunnaki-human hybrid's stepfather, much like Jethro and

102

Joseph. Elizabeth reached an advanced age (though she was not as old as seventy as some sources claim) without conceiving, when an angel, which of course was an Anunnaki, visited the couple and promised them a son.

The scientific Anunnaki system soon impregnated Elizabeth, and the son turned out to be John the Baptist, another hybrid with magical qualities. Elizabeth was chosen for the task for her perfect genetic makeup.

She was a direct descendant of Aaron, Miriam's son.

The intricate, thousands of years long records of such relationships, are meticulously kept by the Anunnaki.

And so the pattern returns, again and again, until it culminates in an entire religion, Christianity, based on a little boy named Jesus who is also a hybrid, born of an immaculate conception. And as long as the connection between Anunnaki and humans continue, the pattern will go on.

*** *** ***

104. Ezakarerdi, "E-zakar-erdi "Azakar.Ki" The Inhabitants of Earth
⌘ ⌘ ⌘

Definition:
Term for the "Inhabitants of Earth" as named by the Anunnaki, and mentioned in the Ulemite language in the "Book of Rama-Dosh."
Per contra, extraterrestrials are called Ezakarfalki.
"Inhabitants of Heaven or Sky". The term or phrase "Inhabitants of Earth" refers only to humans, because animals and sea creatures are called Ezbahaiim-erdi.

Ezakarerdi is composed of three words:
1-E (Pronounced Eeh or Ea) means first.
2-Zakar: This is the Akkadian/Sumerian name given to Adam by Enki. The same word is still in use today in Arabic, and it means male. In Arabic, the female is called: Ountha (Oonsa).
The word "Zakar" means:
a-A male, and sometime a stud.
b-To remember.

In Hebrew, "Zakar" also means:
a-To remember (Qal in Hebrew).
b-Be thought of (Niphal in Hebrew).
c-Make remembrance (Hiphil in Hebrew).

There is a very colorful linguistic jurisprudence in the Arabic literature that explains the hidden meaning of the word "Zakar"; Arabs in general believe that man (Male) remembers things, while women generally tend to forget almost everything, thus was born the Arabic name for a woman "Outha or Oonsa", which means literally "To forget!"

105

Outha (Oonsa) either derives from or coincides with the words "Natha", "Nasa", "Al Natha", "Nis-Yan", which all mean the very same thing: Forgetting; to forget, or not to remember. Islamic scholars explain that the faculty of remembering is a sacred duty for the Muslim, because it geared him toward remembering that Allah (God) is the creator.

Coincidently or not, Zakar in Ana'kh (Anunnaki language) and ancient Babylonian-Sumerian means also to remember. Could it be a hint or an indication for Adam's duty of remembering Enki, his creator?

3-Erdi means planet Earth. Erdi was transformed by scribes into Ki in the Akkadian, Sumerian and Babylonian epics.

From Erd, derived:

a-The Sumerian Ersetu and Erdsetu,

b-The Arabic Ard,

c-The Hebrew Eretz.

All sharing the same meaning: Earth; land.

Thus the word Ezakarerdi means verbatim: The first man (Or Created one) of Earth or the first man on Earth, or simply, the Earth-Man. In other word, the terrestrial human.

*** *** ***

105. Anunnaki's fertilization light "Fana.Ri"
⌘⌘⌘

I. Definition and introduction:

Name for a beam, or an Anunnaki' sort of light used in the fertilization and genetic reproduction process of the Anunnaki.

According to Ulema Maximillien de Lafayette, Anunnaki reproduction is done by technology, involving the light passing through the woman's body until it reaches her ovaries and fertilizes her eggs.

Sex and reproduction are two separate functions. Anunnaki reproduction is done by technology, involving the light passing through the woman's body until it reaches her ovaries and fertilizes her eggs.

The eggs go into a tube. The woman is lying on a white table for this procedure, surrounded by female medical personnel.

If performed by uncaring aliens (such as the grays and the reptilians) it is unpleasant and even can be painful, which has given rise to the abductee's stories of suffering. However, not all aliens are created equal.

The Anunnaki, which are a very compassionate race, are very gentle and the procedure is harmless. Apparently, the Anunnaki version of sex is much more enjoyable for both genders.

It involves an emanation of light from both participants. The light mingles and the result is a joy that is at the same time physical and spiritual. The Anunnaki do not have genitals the way we do.

107

As a hybrid becomes more and more Anunnaki, he/she loses the sexual organs and becomes physically like the Anunnaki. The hybrid welcomes the changes and feels that he/she has gained a lot through the transformation. The Anunnaki mate for life, like ducks. They don't even understand the concept of infidelity, and don't have a word for cheating, mistress, extramarital affairs, etc. in their language.

Like many extraterrestrials, the Anunnaki do not have genital organs, but a lower level of aliens who inhabit the lowest interdimensional zone and aliens-hybrids living on earth do.

The stories of the abductees who claim to have had sex with Anunnaki are to be disregarded. Those stories are pure fiction.

II. Summary of fertilization and reproduction:

- 1-Aliens reproduce in laboratories.
- 2-Aliens do not practice sex at all.
- 3-Aliens fertilize "each other" and keep the molecules (not eggs or sperms,) in containers at a very specific temperature and following well-defined fertilization-reproduction specs.
- 4-Alien babies are retrieved from the containers after 6 months.
- 5-The following month, the mother begins to assume her duty as a mother.
- 6-Alien mothers do not breast-feed their babies, because they do not have breast, nor do they produce milk to feed their babies.
- 7-Alien babies are nourished by a "light conduit."
- 8-Human sperm or eggs are useless to extraterrestrials of the higher dimension.
- 9-Extraterrestrials are extremely advanced in technology and medicine. Consequently, they do not need any part, organ, liquid or cell from the human body to create their own babies.
- 10-However, there are several aliens who live in lower dimensions and zones who did operate on abductees for other reasons – multiple reasons and purposes – some are genetic, others pure experimental.

106. The third level of an Anunnaki-Ulema's Training "Hugari-Darja" "Baa'-La Guri Darja"
⌘⌘⌘

I. Definition
II. Composition of the three worlds or universes
III. The concept in ancient religions

I. Definition:

Expression referring to the third level of an Anunnaki-Ulema's training. More precisely, it is the code or the name given to the Anunnaki-Ulema "Third Degree of Initiation."

Basically, it is an orientation program leading to the activation of the "Conduit", this enigmatic invisible cell genetically created by the Anunnaki, and installed in the brains of early humans at the beginning of time.

The Conduit, once activated can produce supernatural powers and faculties. A student who has completed the Hugari-Darja becomes capable of mastering extraordinary deeds, such as, to name a few:

- a-teleportation,
- materialization,
- dematerialization,
- levitation,
- telekinesis,
- cells' self-reproduction,

- entering a parallel dimension,
- reading others' thoughts, so on.

It was reported by Ulema Naphtali ben Yacob, that the Tibetan "Third Eye" concept originated from the Anunnaki's Hugari-Darja.

Cheik Allamah Sadek Zukri bin Abi Sufian claimed that the ancient practice of "Firasah", which is the technique of reading others' minds, and guessing allies and foes intentions, was inspired by the Anunnaki Hugari-Darja.

In this orientation program, a person is taught secrets about the basic structures of three worlds that are closely associated with humans.

Thus, the term "Third Degree" or "Third Level" is metaphorically used. It is closely related to the "architectural foundation" of the cosmos, and which is the secret origin of the esoteric concept of "Sacred Geometry", and the Masonic third degree.

II. Composition of the three worlds or universes:

These three worlds are:

- **1**-Physical world or physical plane. It consists of our solar system, and similar planetary systems, anywhere in the universe, with identical and/or different laws of physics.

- **2**-The parallel world or parallel dimension. It consists of a multitude of universes and worlds.

- **3**-The Anunnaki Shama. It consists of the highest plane of the cosmos, which is the sphere of the Creation of the universes and all life-forms. From this sphere was created the Big Bang of the universe.

III. The concept in ancient religions:

This concept is found in numerous metaphysical and esoteric teachings and practices. In fact, one form, and/or a similar concept of this Anunnaki Hugari-Darja have existed in ancient religions.

Every nation had its exoteric and esoteric religion, the one for the masses, the other for the learned and elect.

For example:

- **a**-The Hindus had three degrees with several sub-degrees.
- **b**-The Egyptians had also three preliminary degrees, personified under the "three guardians of the fire."
- **c**-The Chinese had their most ancient Triad Society.
- **d**-The Tibetans have to this day their "Triple Step", which was symbolized in the Vedas by the three strides of Vishnu. Everywhere, antiquity shows an unbounded reverence for the Triad and Triangle—the first geometrical figure.
- **e**-The old Babylonians had their three stages of initiation into the priesthood, which was then esoteric knowledge.
- **f**-The Jews, the Kabbalists and mystics borrowed them from the Chaldees, and the Chaldees borrowed this mystic knowledge from members of the "Fish Brotherhood", governed by the Anunnaki-Ulema.
- **g**-And later on, the Christian Church borrowed this from the Jews, who themselves took it from a much ancient Babylonian belief, always influenced by the Anunnaki-Ulema.

*** *** ***

Immortality in Ana'kh Literature
⌘ ⌘ ⌘

I. Did the Anunnaki kings on earth seek immortality?
II. Gilgamesh's visits to Baalbeck and Al Arz
III. Gilgamesh and his search for immortality
IV. What did the Anunnaki's leaders tell the Ulema about our humans' longevity on earth and immortality in the other world?
V. The Ulema continued: "You can live 10,000 years on earth, if you comply with the rules of the Ulema's

Immortality in Ana'kh Literature
⌘ ⌘ ⌘

I. Did the Anunnaki kings on earth seek immortality?
II. Gilgamesh's visits to Baalbeck and Al Arz
III. Gilgamesh and his search for immortality
IV. What did the Anunnaki's leaders tell the Ulema about our humans' longevity on earth and immortality in the other world?
V. The Ulema continued: "You can live 10,000 years on earth, if you comply with the rules of the Ulema's

I. Did the Anunnaki kings on earth seek immortality?
A student asked an Ulema: Did the Anunnaki kings on earth seek immortality? The Ulema replied verbatim (As is and unedited): The answer is yes.

Here is the story...
There is one place on earth, the Ulema consider as the ultimate "terminal" of the Anunnaki; a sort of a Ba'ab from which a person enters or exits a physical dimension. Thousands of years ago, and long before the Sumerians established their kingdom in Iraq, and interacted with the Anunnaki, and many many centuries before the human race in any region of the world learned about God or Gods, the Anunnaki landed on that very special place and revealed to its inhabitants many secrets, including teleportation, psychic healings, and the divine nature of the supreme beings (God, creators).
That place served them as a landing and a launching post.
It still exists today; it is the ancient city of Baalbeck in Lebanon.
Baalbeck was mentioned in many Sumerian, Babylonian and Persians epics, texts, and tablets. Baalbeck was the rendezvous and favorite sacred place of the kings and deities of Sumer, Babylon and Egypt, because it was the first city established by the Anunnaki on earth.

115

The legendary king of Uruk, the Anunnaki king-god Gilgamesh visited Baalbeck many times, and worshiped there. He worshiped higher gods.
Why did he go to Baalbeck?
What did he expect to find?
The answer will astonish you.

II. Gilgamesh's visits to Baalbeck and Al Arz:

Gilgamesh hoped to acquire immortality and extra supernatural powers from the gods who lived in Baalbeck.
The gods welcomed Gilgamesh and told him Baalbeck is the entrance to the other world...to the primordial sphere that created earth and the human race: Ashtari, also known to humans as Nibiru.

The gods of Baalbeck directed Gilgamesh to the secret celestial Ba'ab of his ancestors the Anunnaki; from that Ba'ab (Exit), Gilgamesh could reach Ashtari in a blink of an eye.
The Gods also told Gilgamesh that as soon as he enters the Ba'ab he will become immortal, but he should continue his journey to Ashtari (Nibiru), and Gilgamesh did.
And Gilgamesh asked the gods:

"Will the Ba'ab take me directly to Ashtari,
so I would reach immortality?"
and the gods answered:

"Eventually, but first, you must make a short stop
at the spring of immortality Al Arz,
a sacred region of your ancestors, the Anunnaki..."
And Gilgamesh asked again:
"Where is Al Arz?"
And the gods answered:
"Not very far from here...
it is the highest and mightiest mountain in Phoenicia,
where the Anunnaki your ancestors planted the cedar trees...
and on the top of the mountains you will land
for a short time where you will clean
your thoughts and mind...
and you will stroll under the branches of the cedar trees...and
short after you will continue your journey to Ashtari...

and immortality you shall have..."

Al Arz is a mountainous region in Lebanon (Ancient Phoenicia), where the Biblical cedar trees grow...the same trees King Hiram of Tyre used to build the Temple of Solomon, and King Tut An'k Amoon used to decorate his palaces. Also Noah's boat was built with Phoenicia's Arz trees.

Al Arz means cedar trees in Arabic and modern Phoenician. It is derived from the Anunnaki's language "An'k". The Sumerians believed that Al Arz and Baalbeck are the holy cities where immortality lives as disguised gods on planet earth. This is why Gilgamesh traveled to these two old Anunnaki-Phoenician cities.

III. Gilgamesh and his search for immortality:

The poems "Gilgamesh and Huwawa", "Gilgamesh and the Bull of Heaven", "Gilgamesh and Agga of Kish", "Gilgamesh, Enkidu, and the Nether World", and "The Death of Gilgamesh", relate various incidents and adventures in his life, and his obsession with immortality.
The epic appears on 12 clay tablets found at the site of the ancient Assyrian city of Nineveh. The tablets came from the legendary library of King Ashurbanipal, (685 B.C.-627 B.C.), son of Esarhaddon, and the last great king of the Neo-Assyrian Empire.

IV. What did the Anunnaki's leaders tell the Ulema about our humans' longevity on earth and immortality in the other world?
Because the teachings of the Ulema are not based on esoterism, but rather on the Anunnaki's metaphysics and science, the immortality of Man is no longer a speculative matter for the Ulema and their initiated students.
However, many of the Ulema's students (Almost 99% of novices were initiated, and eventually became enlightened) in their initial training stage were perplexed by the idea, and during the early stage of their training asked so many questions about Man's longevity, immortality and the after-life.

A student asked an Ulema:

Can we live for ever with the angels in heaven?
Are humans immortal since we are the creation of God who is immortal himself?

Note: It is obvious that that question was asked by a student who has not yet received advanced training or attended readings of a high level, because he used words and sentences such as "creation of God" and "angels in heaven".
The Anunnaki and the Ulema do not believe in the same God we worship and fear; the Judeo-Christian-Muslim God, that is, because they know the origin, nature and name of the person, entity or supreme being who introduced himself as God to Abraham, and the early Jewish prophets.

Al Mutawalli said: "That god was a fake god. He was simply one of the early royal leaders of the Anunnaki who had territorial ambitions...Some believe he was Enki, others believe he was Enlil, and eventually, one of them became Yahweh.

Abraham, and before him, his father Terah thought that the Anunnaki Sinhar was a creator, a god and ruler of heavens and earth, while in fact, he was a military Anunnaki commander in Sumer..."

V. The Ulema continued: "You can live 10,000 years on earth, if you comply with the rules of the Ulema.

Excerpts from the answer of the honorable Ulema Rama Nabih:
- The physical cannot enter the non-physical.
- The non-physical can enter the physical and gives it life for thousands of years.
- All humans on this Earth are mortal in this sphere.
- In the future (Used as a terrestrial term), many will live thousands of years.
- The An.NA.Ki (Anunnaki) told us that some Bashar (Humans) will live for 10,000 years on planet earth. This is the maximum of their life-span in this physical dimension.
- Their longevity was written originally as "Shou' LA" (First sparkle of life).

118

Note: Today, we call it DNA. A modern Ulema rephrased the statement as follows: The longevity of a person is always decided upon, the moment he is born. The number of years he will live on earth is already written in his DNA.

- Man will be able to live up to 10,000 years if he complies with the Ulema's Nizam.
- The Nizam requires many things.

The most important requirements/prerequisites (Moutawajibaat, Shou-Rout) to acquire longevity or immortality are:

- **a**-Activation of the Conduit.
- **b**-Synchronization with the Double.
- **c**-Mental access to the Madkhal.

The most important rules (Kawa-ed, Ousool) to acquire longevity or immortality are:

- **a**-Abstinence from eating flesh (Meaning meat).
- **b**-Abstinence from using hallucinatory drugs.
- **c**-Total purification of the Mind. Not even harmful thoughts and intentions are allowed. The ill-thoughts and selfish desires are as bad as ill-deeds and actions.
- **d**-You live by thought and action. Both must remain pure and honorable.
- **e**-Once a year, you must share one monthly earning with the poor and the needy.
- **f**-Completion of the three levels of your training, and receiving the "Barakaat" (Blessings) of the initiation.
- **g**-Compliance with the teachings of the Book (Book of Rama-Dosh).

The honorable teacher added:
- Nothing is immortal on Earth. However nothing is lost for ever on Earth. Everything is re-transformed into something else, and takes on different shapes, forms, meanings, values and dimensions...
- The longevity of Bashar (Human beings) will begin at the end of the year 2022, once the contamination of the

human race has been eliminated by the Anunnaki during that year.
- In other dimensions, starting with the Fifth dimension, the Mind as you new form of existence in the universe acquires limited immortality.

*** *** ***

18. UFOs, USOs
⌘ ⌘ ⌘

I. How Important UFOs are to the Ulema?
I. The Ulema Views and statements
a- Ulema Sorenztein said
b- Ulema Oppenheimer said
II. UFOs' and USOs' Information and details given by Ambar
Anati and Ulema

I. How Important UFOs are to the Ulema?
The Ulema Views and Statements:

a- Ulema Sorenztein said:

Modern Ulema Sorenztein who lived in Lower Manhattan, New York City said verbatim: "UFOs are extremely important to Westerners. In fact, they constitute the very fabric of ufology. Without those unidentified flying objects, Western ufology will cease to exist. To Easterners, UFOs are important too, especially to those persons who have read UFOs books published in Europe and in the United States.

To the Ulema, Mounawarin, Illuminated "Enlightened" teachers, and our students, UFOs are not as important, because they were and remain simply very advanced types and categories of transportation vehicles, despite their extraordinary maneuvers and astonishing capabilities.

Simply put, they are flying machines. Nothing more, and nothing less. They are simply spaceships.

What is paramount and extremely important to us, is not the technical and scientific aspects of these flying machines, but their origin, the nature of their pilots, the reasons for being here, the reasons for visiting Earth, and the technology they could or would offer to the human race.

But the technology part is a minor consideration.

Although, they have astonished us at so many levels, the UFOs remain in our minds and study of galactic civilizations, a mode of transportation, very advanced spaceships, and our interest in them ends right there."

b- Ulema Oppenheimer said:

Ulema Oppenheimer, one of the most brilliant minds of our time said verbatim: "We know where they (UFOs) came from. We know their mode of operation.

We are familiar with their characteristics and specs. We have learned a great deal about their different categories, shapes and performances.

They are no longer "Unidentified" to us. Of course, we do teach our students many courses (Kira'at and Lectures) on the technical aspect of these spaceships, however, the study of the UFOS as striking spaceships is minimal in comparison to more important subjects such as galactic civilizations, beginning of time and space, origin of the human race, and our relationships with very advanced beings from other cosmic systems.

We do explain to our students how they look like, how they travel the cosmos, etc., and the story ends here. To us, the UFOs occupy a small part of our concerns and study. In fact, at one time, and soon or later, students become very bored with Kira'at on UFOs as flying saucers.

Yet, in the West, UFOs have preoccupied the minds of the masses for years, and why is that? There are many reasons, but the most important ones are, to name a few:

- a-The theories, allegations, and religious interpretations about UFOs;
- b-The relation between UFOs and so-called abductees;
- c-The UFOs "Alien technology" and its implication in the military field;
- d-The UFOs as messengers from highly advanced civilizations coming here to enlighten us and help us reach a higher level of awareness;
- e-Sometimes, just the opposite, when UFOs are seen as a form of Satan's greatest deception.
- f-So on...

Ironically, all these reasons and justifications do not worry us at all. For we know that the truth is quite different, and we have nothing to fear or worry about.

The UFOs are just a piece of metal. You could compare them to a very expensive and well-tailored suit. What really counts is that person who is wearing the suit. The nature and origin of that person, and particularly his/her intentions, level of knowledge, mental and spiritual awareness. In the Western hemisphere, the UFOs are the front line of ufology. If they vanish, and do no longer appear in the skies and on land, ufology and books on UFOs could shrink considerably. In our Kira'at and Dirasat, UFOs are just a footnote."

II. UFOs' and USOs' Information and details given by Ambar Anati and Ulema:

Ambar Anati provided us with an astonishing literature and detailed information about UFOs.

We did mention Anati's reports and elaborate description of these machines in previous books.

Herewith, is a selection of the most important details and depictions given to us by Ambar Anati; a sort of summary, a synopsis.

Ambar Anati and other Ulema have said, verbatim, and in no particular order or sequence:

- **1**-Extraterrestrial spaceships rarely crash.
- **2**-If damaged, they are repaired instantly if accompanied by a mother-spaceship, or are in the company of a formation.
- **3**-Singular extraterrestrial craft could crash. This occurs when the craft is traveling "alone."
- **4**-The truth is that many extraterrestrial spaceships constantly travel our skies without being noticed by humans, because they are surrounded by an "invisibility shield." Consequently, they remain undetected.
- **5**-This "invisibility shield" prevents the spacecraft from crashing.
- **6**-The shield could malfunction for numerous reasons, including various electromagnetic discharges, and anti-gravity protection failure. This happens only in our solar system.

- 7-Extraterrestrials do not ask for human help to repair their damaged ships. Always a large mother-ship accompanies their spacecrafts for full technical support.
- 8-The mother-ship is usually very big and enormously spacious. One of its main characteristics is "housing" a large number of spacecrafts.
- 9-So-called damaged spaceships are repaired in hangars located inside the mother-ship, usually found at the lower level of the interior of the mother space-ship. Consequently, damaged spaceships do not need to land be repaired.
- 10-Extraterrestrial damaged ships are never repaired manually. Usually, aliens' spacecrafts are self-repaired above ground while flying.
- 11-A damaged spaceship is never visible to the human eye. And if this incident occurs, the spaceship releases a substance from a section under its "belly". This substance is quickly coagulated as soon as it touched the surface of the earth. And in no time, it disintegrates, leaving no trace.
- 12-It is important and very useful to remember that extraterrestrial spaceships do not fly as wrongly and mistakenly described in the West; they "jump", so to speak for lack of proper terms.
- 13-They reach destinations by traveling through "aerial lines and curves" incomprehensible to humans.
- 14-They belong to an infinite time, rather to a limited space. And when they appear in our skies, they adopt a zigzagging pattern for sharp turns and for speed acceleration.
- 15-If they want to stand still, they hover "motionless" and in complete silence.
- 16-Extraterrestrial spaceships do not use engines, propulsion systems or fuel. Almost all of them use two "devices": 1-An antigravity module for traveling inside the solar system and for landing. 2-A space/time apparatus for galactic travel. The human mind can not understand how these two "devices" work.
- 17-Therefore, humans can not reverse the engineering of extraterrestrial crafts, unless taught by extraterrestrials in person. In addition, humans need very advanced

repair tools, material and spare parts found only outside the solar system. Therefore, no scientist on earth has ever succeeded in deciphering and learning how to fly or repair an extraterrestrial scapeship.

- **18-**The most advanced types of USOs are the crescent-shaped ones. They are smaller than the circular ones, and usually are deployed over mountainous areas. They fly in formations, and are tele-guided by a space-time memory apparatus.

- **19-**The memory apparatus works like a navigation device, but not like a compass, because it does not have altitude, longitude and directions/positions for south, north, west and east.

- **20-**Many of the spaceships we see in our skies were constructed 1,000 years in our future.

- **21-**Many of the spaceships we see in our skies were here on Earth thousands of years before the human race existed on earth.

- **22-**The greater number of spaceships comes from underwater; straight from seas and oceans.

- **23-**UFOs' "invisibility shield" could malfunction for numerous reasons, including various electromagnetic discharges, and anti-gravity protection failure. This happens only in our solar system.

- **24-**Many UFOs (Not all of them) are "timeships." Outside the solar system, time and space are not found or defined separately. In fact, there is no time, and there is no "physically limited" space in the outer galaxy. The human mind is incapable of understanding this, yet! In the future, scientists on earth will eventually figure it out.

- **25-**The creation is made out of millions of dimensions, parallel universes, future universes, humans with multiple doubles sharing an infinite space.

- **26-**As soon as a spaceship, or any object for that matter, exceeds the speed of light, it finds itself moving back in time. It might sound strange how a vehicle that goes faster than 186,000 miles/second can travel backward in time, but it is perfectly logical.

- **27-**An increase of motion slows time.

- **28-**When a spaceship approaches C, the speed of light, time slows down until at C time stops.
- **29-**It is obvious that a spaceship cannot go faster if time has already stopped. However, when the vehicle crosses the light barrier, time goes even slower on the other side, thus the spaceship has entered into the realms of negative time and find itself traveling back in time.
- **30-**The crew and pilots of UFOs are not of a human origin.
- **31-**UFOs/USOs coming out of the oceans are also piloted by intraterrestrials, some are totally aliens (Different DNA), others are hybrids.
- **32-**The UFOs show few, if any, signs or indications of wishing to communicate with the human race.
- **33-**Supporters of the "UFO future human vehicle theory" will object by saying that communication with the present human species is not allowed, because it would interfere with our timeline. This is not correct.
- **34-**Few UFOs have crashed on Earth; some have been intentionally shot down. But these cases are extremely rare.
- **35-**Some extraterrestrial and intraterrestrial spaceships and their crew (Dead or alive) have been secretly recovered by the military in the United States and Russia.
- **36-**UFOs' pilots belong to many and different races and species.
- **37-**One of them might be the future human race.
- **38-**The construction of a time machine such as an UFO is beyond our reach for a very long time.
- **39-**Even if we succeed in building a time travel machine, we can only go back in time to a period when the time vehicle was created.
- **40-**If a UFO (Time machine) is constructed in 2,100, it cannot be used to transport you back to the year of 1,100.
- **41-**Many UFOs are experimental government crafts.
- **42-**Many UFOs have alien bases on Earth.
- **43-**None of these UFOs are piloted by Anunnaki.

- **44**-Almost 90% of UFOs you see in the skies are the product of a non-human race, sharing our habitat. They are not totally extraterrestrials.
- **45**-Some UFOs and USOs come from multiple alternate dimensions, a parallel sphere, where each 3 dimensional space is separated by dark matter. But the majority is from here; Earth!
- **46**-Some UFOs are robotic probes, sent many years ago from other stars. These cases are very rare.
- **47**-Water UFOs are the sky and land UFOs. The UFOs you see in your skies come from bases underwater on planet Earth. They are manufactured by extraterrestrials living on earth and underwater. They are not a product of the Anunnaki, but the Anunnaki know a lot about these mysterious machines.
- **48**-It is extremely rare to sight UFOs coming from outer space, from and beyond our galaxy. It did happen a few times, but what we usually see are machines from our future.
- **49**-Water UFOs are ageless. They were manufactured millions of years in our past which becomes our future
- **50**-UFOs do not suffer from metal fatigue, because the metal used has its own memory. And that memory reconditions it.
- **60**-Underwater, the color of the USOs changes according to the environment, the density of the water and the aquatic conditions of the milieu.
- **61**-Many UFOs/USOs are surrounded by an anti-gravity perimeter and protected by a shield non visible to the naked eye and to the most sophisticated radar on earth, or any observation-reconnaissance device.
- **62**-There are no linear or trajectories information on board of the USOs, because, in reality, USOs/UFOs do not fly; they change positions, in other words, they "jump from one pocket to another." Pocket means the "rendez-vous" of entry and exist of "anything" in the time-space opening that occurs when an object, even a thought escapes the "measurable." To millions of civilizations in the universe, it is a basic scientific knowledge.

On Earth, scientists tried to explain this "rendez-vous" of time and space as "parallel universes", "multiple universes", "higher dimensions, "the M Theory", "future universes", "wormholes", "time warp", "black holes attracting forces", "reversed time", "time travel", "space-time travel", "black holes absorbing energy". This is the best they could do.

- **63**-The "rendez-vous" is infinitesimally small universe molecule capable of expanding into the shape of a tube, where physical energy and earthy laws of physics cease to exist; a UFO-USO can penetrate this molecule (call it tube if you want or even wormhole) and reach multiple destinations faster than a beam of light.

- **64**-The opening of the "rendez-vous" exists only for a fraction of time. We used "time" because that is the only measurement you can understand for now.

- **65**-The speed of light at that fraction of time is created by a negative energy that distorts the fabric of space; hence, time as you know it and understand it ceases to exist.

- **66**-Extraterrestrials (as you call them) are capable of opening this "rendez-vous", travel through it, and exit before it closes up or collapses on its own time-space dimensions.

- **67**-USOs penetrate the water without creating splashes, and come out of the water "without dripping."

- **68**-Before the USO touches the surface of the water, it shoots an invisible beam that separates the lower part of the space ship from the water, thus creating a "vacuum space" similar to an air pocket.

- **69**-USOs do not navigate waters. They "slide" through an "air-vacuumed tunnel" called "Plasma Corridor", that opens up right in front of them, and closes behind them as soon as they cross it.

- **70**-Corridor plasma is a term used to refer to underwater tunnels and passages created and operated by aliens to navigate the oceans. By using these plasma tunnels, UFOs can accomplish extraordinary tasks, such as, to name a few:
- **a**-Reach an astonishing speed;
- **b**-Avoid sonar detection;
- **c**-Remain undetected by spy satellites;

- **d**-Enter and exist underwater bases.

The corridor plasma is movable and mobile, meaning that aliens can place the underwater tunnels, and displace them according to their needs, and "navigation chart". The tunnels extend to thousands of miles underwater, and serve as a web network for several alien underwater bases around the globe.

Some of these bases are located in the Bahamas, the Japanese "Dragon Triangle", the north side of the so-called Bermuda Triangle, Alaska, and Florida.

- **71**-The water never touches the body of the scapeship.
- **72**-Many of the spaceships we see in our skies were here on earth thousands of years before the human race existed on earth.
- **73**-The greater number of spaceships (UFOs, USOs) comes from underwater; straight from seas and oceans.

*** *** ***

109. Sinhar Ambar Anati (The Anunnaki-Hybrid Wife) said⌘⌘⌘

109. Sinhar Ambar Anati (The Anunnaki-Hybrid Wife) said
⌘⌘⌘

I. Ambar-Anati "Anbar-Anati
II. Signs and Marks of the Presence of the Anunnaki
III. Ascension and Cleansing of humans in 2022
IV. In 2022, On Earth, All "Ba'abs" (Stargates) Will Open Up
V. Where are located these Ba'abs?
VI. Anunnaki vis-à-vis other alien races
VII. Anunnaki's characteristics

I. Ambar-Anati "Anbar-Anati":
Ana'kh/Akkadian/Sumerian.
Name of an Anunnaki's hybrid-female leader who has contacted humans in recent time, being herself an Anunnaki-human hybrid. Composed from:
a-An (High; heaven; sky).
b-Bar or Bau (Sky; high; bright).
c-Ana (First; original).
d-Ti (A rib; origin of creation).
General meaning: Leader descendant from the great mother of Ea, reigning over the heavens and light.

Herewith, you will find some of her most boggling statements, explications and revelations on a wide variety of subject, ranging from bending time-space to the Anunnaki's return in 2022:

133

- Bending time and space is not enough. To traverse the whole universe. It could be done through this method, but the most effective way, is the Anunnaki's way: Through light, because light does not have mass. It seems crazy stuff, but some American scientists are already exploring this possibility.
- The United States government is fully aware of the return of the Anunnaki. The American military and top scientists were warned and advised about a major confrontation with extraterrestrials. The aggressive clash is inevitable.
- So far, the American government has kept everything under a heavy lid. They do not know how to deal with the issue or how to present and explain the situation to us.
- The Anunnaki will return in 2022, and they will be using three types of spaceships:

a-Huge and massive mother spaceship;
b-Conventional circular spaceships for transport;
c-Crescent-shaped spaceships for "Ascension" following the cleansing of humans who were contaminated by earth elements and Grays' DNA genetic manipulations.
In other words, each type of these spaceships will carry and execute very specific tasks and functions.

- The triangular spaceships will be used by Anunnaki's leaders crossing the "Ba'abs".
- Earth's governments and armed forces will not resist. However, because a very large portion of the civilian population will be invaded by waves of hysteria, fear and panic, many individuals and groups will counter-attack the Anunnaki's spaceships, and this will result in mass annihilation of human beings.
- The Anunnaki's "Miraya" shows that those who are more likely to adopt an aggressive attitude toward the Anunnaki are American citizens. The "Miraya" also revealed and displayed the locations and states where such unfortunate actions will take place. Sinhar Ambar Anati did indeed give us the exact locations and name of the states located in the United States.

II. Signs and Marks of the Presence of the Anunnaki:

- One of the Anunnaki trademarks is leaving behind them signs and traces of their presence, and or past existence in time and space (Instead of place).
- For instance, the Anunnaki left on earth huge "marks" and "traces" such as the "Hajarat Al Houbla" in Baalbeck, Phoenicia (Modern day Lebanon), and Melkart healing center in Arwad, Syria, and the "Face on Mars."
- They will do the same thing in 2022 to mark the end of the past civilization of humans on earth, and the beginning of a new human race on earth. Similar "mark" will be "engraved" in the parallel dimension which is not so far away from the third and fourth dimensions.

III. Ascension and Cleansing of humans in 2022:

- In 2022, after the "Ascension" and "Cleansing" of humans (Contaminated by Earth's and Grays), all of you...those who are saved by the Anunnaki will see with their own eyes, huge projections of holographic illustrations of the entire history of humanity from day one till 2022.
- And those who are teleported to higher dimensions, will be able to see and foresee forthcoming events in the future and occurrences on other planets and in parallel (Multiple) dimensions and universes.
- For instance, a saved human being will be able to revisit the past as it occurred on planet Earth, and in other worlds. Also, the saved human being will acquire additional mental faculty that enable him/her to enter other dimensions and witness the birth of new worlds, dimensions, the creation of new stars, planets, and even galaxies.
- If interested in religions (They will be abolished in 2022), a saved human being can tune to spatial channels and frequencies that "reconstitute" all the anterior and antecedent events and lifespan of a particular religious figure and/or era.
- Those saved beings will be able to see and understand the entire phases of the lives of these religious figures.

- The holographic illustration (Screens) will also project sounds captured and recorded from the past.
- In 2022, the Anunnaki will activate 13 chakras in the human body. Saved humans will be able to cure themselves from all diseases and sicknesses.
- They will become immune to all viruses and bacteria...and their newly acquired bodies will be in perfect harmony with the rhythms of the physical and non-physical dimensions.
- In other words, the new human race will no longer suffer from fatigue, physical pain, illness and diseases.
- All saved human beings will become prettier, healthier and much much younger..."
- Now, your mind will not allow you to fully understand what is going to happen to the minds and bodies of human beings in 2022. "Much much younger" means that the newly created human body will not deteriorate as fast as it does now...and the saved ones will not age at all.
- It is not immortality I am talking about, but a state, where the new pure human being can "Stop the process of aging"...another wonderful thing is going to happen to all these good people...they will be able to go back in time and choose what age they want to stay in...for instance, should a pure human being (For example, a person who was saved by the Anunnaki at age 70 in 2022) decide to be 21 year old again, well...he/she will be 21 year old again...and will stay 21 for ever."
- In 2022, the Anunnaki will create new categories of human bodies.
- In 2022, the Anunnaki will give the saved humans, new bodies and new minds' (Mental/Intellectual) faculties and will be classified as:

a-Earth born;
b-Heaven born;
c-God or spirit born.

IV. In 2022, On Earth, All "Ba'abs" (Stargates) Will Open Up

- In 2022, and on Earth, all the "Ba'abs" (Anunnaki's Stargates" will open up.
- These "Ba'abs" are not in outer space.
- They are so close to cities and places on earth. The Anunnaki will open these gates right before your eyes, and become the stargates.
- The Ba'abs existed on the cosmos map since its creation. Some are in the Milky Way, others in far distant galaxies, and many are here so close to your cities, schools, streets and homes...they are almost everywhere around the Earth.
- According to the Ulema, and as explained in the "Book of Rama-Dosh", the difference between artificial stargates and Anunnaki's "Ba'abs" is this:
- Artificial gates take people and spaceships in a sort of time-space wrap-lines-tunnels, but do not take the travelers to a higher dimension of cosmic awareness and a universal enlightenment.
- Artificial stargates do not provide the travelers with mental, or intellectual ascension.
- They are purely mechanic and scientific.
- The Anunnaki's stargates give the travelers full access to the map of the universe, ultra-dimensions, multiple universes, parallel words, and above all the "Higher Dimensions", where the "Ultimate Knowledge", and "Global Salvation" reside."
- Traversing the artificial stargates requires physical sparceships, whether they are created by humans or non-humans.
- Entering, exiting and going through the Anunnaki's "Ba'abs" do not require spaceships on an exclusive basis. Spaceships of Anunnaki and advanced extraterrestrial races do not travel or navigate the skies, they "quantum jump" from one dimension to another.

V. Where are located these Ba'abs?

Ulema Maximillien de Lafayette said, some twenty five years ago, I asked two Ulema, Dr. Farid Tayara and Cheik Al Mutawali: "Where are these Ba'abs?" Dr. Tayara said, verbatim: "They are everywhere. Ba'abs are called Ba'abs because they are so close to us, just like the front door of your house...if the main entrance (Door) to your home is far away from your home, then it is worthless. A front door must be at the very entrance of your residence...same thing for the Ba'abs and their locations vis-a-vis Earth.

Madinat AlKahira "City of Cairo" for instance has 7 huge Ba'abs (Stargates)...there is a major major stargate in Baalbeck, another major one in Petra (Jordan), and a very big one in Languedoc (France).
Honorable Cheik Al Mutawali said: "Sometimes, some Ba'abs have a double entrance with one single exit. The double entrance consists of two narrow parallel layers so thin, and so close to each other like two Chinese rice paper sheets...you can't separate them or take them apart, yet, fleet of huge spaceships of several miles in length will go through like a feather..."
These were his own words...
Sinhar Ambar Anati said: "They are everywhere...absolutely everywhere, to name a few:
- New York City has 14 of them,
- Nevada 7,
- Arizona 6,
- Texas 6,
- Washington DC/Virginia Metropolitan area has 5,
- California has 7,
- Dover in England has one,
- New Delhi has 2,
- Bombay has 3,
- Arwad has one,
- Lima has 2,
- Yucatan has one,
- Tyre has one, etc..."

VI. Anunnaki vis-à-vis other alien races:

Anunnaki look very different from the Zetas and the numerous alien races that have visited the earth. In this context, they never appear or manifest like reptilians or short "Greys".
Sinhar Ambar Anati said:

138

- "It is easy to recognize an Anunaki, because he usually appears like a tall warrior.
- His vest is made out of thin layers of metal called "Handar".
- His wears a long robe "Arbiya" of dark colors.
- Underneath the Arbiya, he wears a sort of pans with wide contour.
- On his wrist, you always notice a navigation tool."

VII. Anunnaki's characteristics:

- The Anunnaki can transmute and manipulate their bodies if needed. This happens very rarely.
- In many instances, they don't need to do so, because they are already known to so many galactic and outer-galactic civilizations, and are seen by inhabitants of millions and millions of stars, planets and moons.
- They are superstars in their own rights.
- The Anunnaki can easily shape-shift themselves. This is necessary for climatic and atmospheric reasons.
- Each planet, star and dimension has its own climate, temperature and atmosphere.
- Consequently the organic-galactic body must adapt to these environment's conditions in order to remain functional.
- When they visit earth or a similar sphere, the Anunnaki slightly change their physical appearances, not so much, because in general, they look like us, except their eyes are much bigger, and their height is far more superior and taller to the size and height of humans.
- Some Anunnaki are 9 foot tall.
- Even their women are extremely tall by human standard. Some women are 8 foot tall.
- When they travel to other planets, minor changes are required. For instance, when they get out of their galaxy and visit the planets Niftar, Marshan-Haloum and Ibra-Anu, they change the color of their skin, and the shape of their hands. On Niftar, inhabitants have grey-blue color skin and 3 fingers in each hand.

110. The Anunnaki-Ulema
⌘⌘⌘

1-Category One: The Noubahari, "Noubarim", "Noubari"
2-Category Two: The Mou-Na.rin "Mounawariin",
"M'Noura-Iin"
3-Category Three: The Gayir-Mirayin "Gayrmirayim"
4-Category Four: The Ari-Siin "Arishim"
History synopsis
5-Anšekadu-ra abra "Anshekadoora" abra

110. The Anunnaki-Ulema
⌘ ⌘ ⌘

1-Category One: The Noubahari, "Noubarim", "Noubari"
2-Category Two: The Mou-Na.rin "Mounawariin", "M'Noura-Iin"
3-Category Three: The Gayir-Mirayin "Gayrmirayim"
4-Category Four: The Ari-Siin "Arishim"
History synopsis
5-Anšekadu-ra abra "Anshekadoora" abra

The Anunnaki Ulema are classified and categorized as follows, by order of importance and hierarchy, starting from the lowest level to the highest one:

1-Category One:
The Noubahari, "Noubarim", "Noubari".
Noubahari is the plural of Noubih.
Noubih is either a noun or an adjective. It means alert, informed, observant, wise, messenger of truth and wisdom.
From Noubih, derived the Sumerian and Akkadian words Nabih or Na. Bih, which means messenger, and the Arabic word Nabih, which means wise, intelligent, and well- informed.
The Noubahari are humans, and they live on earth. Physically, they are not very much different from the rest of us. But on other levels, they are far more superior.

For instance (to name a few):
- **1**-They do not age as rapidly as we do. A seventy year old Ulema looks like a 37 year old man. Ulema Sadik said: "Physically, the Ulema do not look older then 37...and they stay like that for the rest of their lives..."
- **2**-Ulema live longer than ordinary human beings. Their lifespan on earth is approximately 135 years.

143

- **3**-They are vegetarians. Yes, they do drink, but with moderation. Some smoke, but not cigarettes. Their tobacco is made out of aromatic dried fruits.
- **4**-They have an enormous compassion toward animals. They communicate exceptionally well with animals; the majority of animals except crocodiles, snakes, insects carrying bacteria and diseases, and some reptiles species.

Animals sense their presence and welcome them. Ulema have developed a sign language to facilitate their communication with animals. And usually, animals respond in the same manner.

- **5**-Ulema are well-versed in many languages. And they are fond of languages of ancient civilizations, including those of vanished cultures. Ulema learn foreign languages very easily and rapidly. Usually, an Ulema learns a foreign language in less than a week.
- **6**-Ulema can read a voluminous book and memorize it in its entirety in less than three hours.
- **7**-Ulema can foresee the future and predict events to happen in several dimensions, including our own.
- **8**-Ulema are in constant contact with the Guardians.
- **9**-Ulema knowledge in arts, science, history and religions is limitless, etc...

These qualities and gifts allow them to fully understand the human psyche, read our minds, and sympathize with our tastes, needs and aspirations.

They are socially active, however, they do not reveal themselves to the rest of us, nor do they get involved in groups' activities. They dislike organized religions, politics, fanaticism, prejudices, stock markets, financial interests, publicity, vain public debates, egoism, and excessive authority.

It is not so easy to gain membership in their groups and societies. Membership is by invitation only. Membership procedures and initiation process, formalities, and rituals are rigorous. Many applicants have failed because of the tests they had to go through.

⌘⌘⌘

2-Category Two:
The Mou-Na.rin "Mounawariin", "M'Noura-Iin".
Ana'kh/Ulemite.. It means the enlightened ones.
It is either a noun or an adjective.
From the Ana'kh word Mou-Na.rin, derived the Ulemite term Mounawariin, which literally means people of the light, or more precisely the illuminated ones.
The Mou-Na.rin are humans, and they live on earth.
They are a group of thinkers, philosophers and scientists. They are the custodians of important books and ancient manuscripts about the origin of mankind, the creation of the universe and human races, as well as a multitude of subjects pertaining to vital aspects of humanity, non-terrestrial-intelligent beings, Arwah, and other dimensions that are closely connected to humans, and non-humans.
The Mou-Na.rin can contact non-terrestrial beings and entities via several and multiple techniques and means.
They can read thoughts, foresee future events, and cure people from all sorts of illnesses and diseases.
A group of philologists and linguists of alternative epistemology believe that the Ulemite term Mounawariin means the people who came from the fire, because the Ulemite term is composed of two words: Mouna or Min which means from, and Narin or Nar, which means fire.

Another group of scholars suggests that the term Mounawariin means people who are surrounded with light, especially around the top of their head, similar to the Buddhas, and saints, because the term is composed of two words:

- 1-M, pronounced Meh or Miin, which means from, or came from;
- 2-Noura (Niir in Ana'kh), which literally means light.

It can be found in several languages, including:

- a-Proto-Hebrew/Hebrew with the word Menora, which means many things including light, candle, lamp, candelabra branches.
- b-Proto-Aramaic/Aramaic/Assyrian with the words Nourah, Nour, which mean light, flash of light, brightness.
- c-Arabic with the word Nour, which means light.

- **d**-Ousmani, ancient and contemporary Turkish with the word Nour, which means light.
- **e**-Farsi/Persian with the word Nour, which means light, and specifically heavenly light.
- **f**-Urdu with the word Nour, which means light, and quite often referring to a religious light and spiritual inspiration.

Thus the complete meaning of the term becomes: People of the light. In esoterism, occult, black arts, Freemasonry and ultimate knowledge studies, the word light means ultimate knowledge and enlightenment.
Bodhisattva in Sanskrit. In westernized version (Probably not totally accurate), they are called the Illuminati.

⌘⌘⌘

3-Category Three:
The Gayir-Mirayin "Gayrmirayim": Ana'kh/Ulemite.
It is composed of two words:
- **a**-Gayir or Gayr, which mean without.
- **b**-Mirayin or Mirayim, which mean visible, and/or could be seen.

The general meaning (Verbatim) is: Those who you can't see.
The Gayir-Mirayin are the non-Physical Ulema.
They do not reveal themselves to us. They communicate with the physical Ulema on an exclusive basis through:
- **1**-Secret codes and a visual language.
- **2**-Ectoplasmic apparitions.
- **3**-Transmission of mind.
- **4**-Visitations through Ba'abs.
- **5**-Telepathy triggered by a "Conduit" implanted and activated in the brain' cells. Ordinary human beings are not trained nor prepared to communicate with them. They can't see them, and they can't sense their presence, even though sometimes they are very close to them.

⌘⌘⌘

4-Category Four:
The Ari-Siin "Arishim".

It means the noble and strong guardians or attendants, also the giant spirits or minds of knowledge. It is composed of two words:

- **a**-Ari, which means big; giant; powerful; attendant; guardian; superior; guide;
- **b**-Siin (Also Shi-yin), which means mind; spirit; ultimate level of knowledge and science.

From the Ana'kh Ari, derived:

- **a**-The Sumerian words A-ri, which means giants, Aris, which means a grant, and Arig, which means attendant;
- **b**-The Assyrian words Ari and Aria, which mean giants;
- **c**-The Hebrew word Ari, which means a lion, and the name Ariel, which means the lion of God (Ari=giant, and El=God);
- **d**-The Hittite word Ari, which means long.
- **e**-The Ulemite Ari, which means those who have.

The Ari-Siin live and evolve in various higher physical and non-physical dimensions.
And this includes the physically known universe, and the meta-cosmos (The world Beyond). They are neither human beings, or spirits. They are pure wisdom and energy.

⌘ ⌘ ⌘

5. History Synopsis:

The Ulema group or brotherhood was created during the time of Hiram, the Phoenician King of Tyre and King Solomon's ally.
The group included astronomers, physicians, mathematicians, artists, scientists, spiritual guides, metaphysicists, philosophers, authors, and lecturers from Sumer, Babylon, Assyria, Phoenicia, Syria, Palestine, Israel, Egypt, China, Mongolia, and Greece.
Later on in history, leading figures of the Knights of St John of Malta, The Templars, The Wise Men of Arwad, and Hiram-Grand Orient Masonic Rites' members joined the Ulema group.
People are taught to believe that the world (Seen and unseen) consists of a physical life on Earth, and a spiritual life after death. The Ulema's views are different. According to The Book of Sun of

the Great Knowledge, the world or universe usually referred to as "Wu-Jud" contains more than a physical life and a spiritual life. Wu Jud consists of 11 dimensions. Humans are aware of three dimensions only.

Some have learned about additional dimensions through theoretical quantum physics, but their knowledge of these extra-dimensions is minimal, or simply theoretical.

The fourth dimension is the one that exists in the next life. That is the limit of Man's understanding and interpretation of the world; the physical and non-physical (spiritual). To the Ulema, life, the world, including human existence go beyond the fourth dimension. For instance, the "Guardians" live in the fourth, fifth, and sixth dimensions. In the seventh and eight dimension, live the "Ultimate Ones", and so on...

Thus, the "Guardians" who live in higher dimensions are noble entities who communicate with chosen human beings and enlightened teachers for various reasons and purposes.

The "Guardians" are not physical beings, however, they can manifest to us in any shape or form using a "Plasmic" organism or substance that the human mind cannot comprehend. The Ulema receive knowledge and guidance from the "Guardians".

The Ulema group was also called the "Society of the Book of Rama-Dosh".

The Ulema do not discuss religions.

They are the custodians and guardians of mind-bending scrolls, manuscripts and books such as:

- ❖ 1-The Book of Rama-Dosh; main topic: The origin of mankind, and how various extraterrestrial races genetically created the human race.
- ❖ 2-Chams El Maaref Al Koubra (The Sun of the Great Knowledge); main topic: The study of superior beings who live in higher physical and non physical dimensions, and who are watching over us.
- ❖ 3-Al Hak (Justice and Truth); main topic: Laws that allow mankind to live righteously on earth and allow human beings to prepare themselves for the next life.

Guidance for the next journey is provided in metaphors and parables.

⌘⌘⌘

6. Anšekadu-ra abra "Anshekadoora" abra:

Expression. An Anunnaki word, meaning learning by traveling or traversing other dimensions. Composed of four words:

- **a**-Anše, which means magnificent.
- **b**-Kadu, which means ability, or to be able.
- **c**-Ra, which means heavenly; godly.
- **d**-Abra, which means to cross over; to traverse.

Ulema Shimon Naphtali Ben Yacob explained that the Anunnaki –Ulema have acquired an enormous amount of knowledge by entering different dimensions, and visiting a multitude of universes.

These dimensions are sometimes called parallel universes, future universes, and vibrational spheres.

He said: "Some are physical/organic, others are purely mental. There are billions and billions of universes. Some are inhabited by beings, super-beings, multi-vibrational beings, and even negative entities known to mankind as demons and evils.

Planet earth is considered the lowest organic and human life-form in the universe. The human beings are the less developed living entities, both mentally and spiritually." Anšekadu-ra abra also means Anunnaki's branching out and changing individuality in multiple universes.

To understand the concept, consider this scenario said Ulema Mordachai Ben Zvi: "Let say, you wish if you could do something differently in your life, something like changing the past, changing a major life decision you have made some years ago, like perhaps, going back in time to a point before you have made a bad decision. Or for instance, you wish if you could do something really good by changing an entire event that has happened in your past.

In the Anunnaki-Ulema's case, they have the solutions for these dilemmas. They can go back and forth in time and space, including the past, the future, and meta-future. An Anunnaki can split himself/herself in two, three, or more if necessary, and move on to a universe that is very much like the one they live on (Nibiru), or totally different. There are so many universes, and some of them do not resemble Nibiru at all.

If an Anunnaki and/orUlema wishes to branch out and move on, he/she must study the matter very carefully and make the right selection. And the branching, or splitting, results in exact copies of the person of the Anunnaki, both physically and mentally.

At the moment of separation, each separate individual copy of an Anunnaki grows, mentally, in a different direction, follows his or her own free will and decisions, and eventually the two are not exactly alike."

So what do they do, first of all? Ulema Ben Zvi said verbatim: "The old one stays where he/she is and follows his/her old patterns as he/she wishes. The new one might land one minute, or a month, or a year, somewhere, some place, right before the decision he/she wants to change or avoid.

Let's take this scenario for instance; some 30,000 years ago in his life-span; an Anunnaki male was living a nice life with his wife and family. But he felt that he did not accomplish much, and suddenly he wanted to be more active in the development of the universe; a change caused by witnessing a horrendous event such as a certain group of beings in his galaxy destroying an entire civilization, and killing millions of the inhabitants, in order to take over their planet for various purposes.

It happened while an Anunnaki was on a trip, and he actually saw the destruction and actions of war while he was traveling. It was quite traumatic, and he thought, at that moment, that he must be active in preventing such events from occurring again, ever.

So, he went back in time to be in a spot to prevent these fateful events from happening again. There, in that new dimension, the Anunnaki leaves his former self (A copy of himself) as a guardian and a protector.

The other copy (Perhaps one of the original ones) is still on Nibiru. The branching out phenomenon occurred in one of the designated locals of the Anunnaki Hall of Records, also called in terrestrial term Akashic Records Hall."

Ulema Openheimer said: "For the Anunnaki fellow there is no problem or any difficulty in doing that. He/she will go back in time and space and change the whole event. This means that this particular event no longer exists in a chronological order. This also means that the event has been erased, because the Anunnaki can de-fragment the molecules, the substance, the vibrations and the fabric of time, but necessarily space.

In other words, that event never happened in one dimension, but it is very possible, that it might still exist in another world. You could consider part of the cosmos are an assemblage of several layers of universes, each one on the top of the other, and sometimes parallel to each other.

When the Anunnaki traverses more than two layers, we call this Anšekadu-ra abra." Can a human being traverse multiple layers of time and space? Yes, said Ulema Ben Zvi.

He added: "However, the human being will be facing a series of problems.

For instance:

1-Case one:

Although, he/she may cross over and enters another dimension, and succeeds in altering, changing or even erasing a past event, the human being might get stuck in that dimension, and remains there for ever.

In this condition, he/she is transformed into a new person without an identity or a past.

A brand new person who is out-placed, without a job, without a residence, without credentials, and without social or professional context. It would become very difficult for that person to make a living. How the others would look upon him?

A person from the past?

A person from the future?

It is not an easy situation.

2-Case two:

Because everything in the world is duplicated ad infinitum in many universes, only one copy of the past event has been altered.

3-Case three:

What would happen to that person, should he/she decides to return back...to his/her original world?

The real problem here is not how to go back to his/her world and relive an ordinary life, the life he/she had before, but what is going to happen to him/her when he/she leaves the new dimension he/she has entered?

Every time a person enters another dimension, he/she created a new copy of himself/herself, and occupies a new spot on the cosmos net. In our case here, that human being by entering another dimension, he/she has duplicated himself/herself in that

151

new dimension, and returning to Earth, he/she will be facing another copy of himself/herself."
Is this possible?
Quantum physic theorists say yes. And they add that humans can enter and live in multiple universes and acquire new identities, and new copies of themselves.

*** *** ***

111. An-Hayya'h, "A-haYA" "Aelef-hayat"
⌘⌘⌘

I. Definition and introduction
II. An or A or Aelef
a- Babylonian account
b-The Chassidut
c- In Hebrew
d- Jewish teaching
III. Hayya'h, Hawwa and Eve
IV. The Anunnaki's equation
V. The Anunnaki's "Conduit"

111. An-Hayya'h, "A-haYA" "Aelef-hayat"
⌘⌘⌘

I. Definition and introduction:

Ana'kh/Ulemite term. An-Hayya'h could be the most important word in the whole literature of the extraterrestrials and Anunnaki because it deals with:

- 1-The origin of man on earth;
- 2-How humans are connected to the Anunnaki;
- 3-Importance of water vis-à-vis humans and Anunnaki;
- 4-The life of humans;
- 5-Proof that it was a non-terrestrial woman who created man, Adam and the human race via her Anunnaki identity;
- 6-The return of the Anunnaki to earth;
- 7-Humanity salvation, hopes, and a better future for all of us; "a gift from our ancestors and creators, the Anunnaki," said the Ulema. It is extremely difficult to find the proper and accurate word or words in our

155

terrestrial languages and vocabularies. The word "An-Hayya'h" is composed of:

II. An or A or Aelef:

An or A (Pronounced Aa), or Aelef (Pronounced a'leff). It is the same letter in Ana'kh, Akkadian, Canaanite, Babylonian, Assyrian, Ugaritic, Phoenician, Moabite, Siloam, Samaritan, Lachish, Hebrew, Aramaic, Nabataean-Aramaic, Syriac, and Arabic. All these languages are derived from the Ana'kh.

(Note: The early Greeks adopted the Phoenician Alphabet, and the Latin and Cyrillic came from the Greek. The Hebrew, Aramaic and Greek scripts all came from the Phoenician. Arabic and most of Indian scriptures came from the Aramaic.
The entire Western world received its languages from the Phoenicians, the descendants of the Anunnaki.)

An means one or all of the following:
- 1-Beginning
- 2-The very first
- 3-The ultimate
- 4-The origin
- 5-Water.

On earth, this word became Alef in Phoenician, Aramaic, Hebrew, Syriac and Arabic. Alef is the beginning of the alphabet in these languages.
In Latin, it's A, and in Greek is Alpha. In Hebrew, the Aleph consists of two yuds (Pronounced Yood); one Yud is situated to the upper right and the other yud to the lower left. Both Yuds are joined by a diagonal *vav*.
They represent the higher water and the lower water, and between them the heaven. This mystic-kabalistic interpretation was given to us by Rabbi Isaac Luria.
Water is extremely important in all the sacred scriptures, as well as in the vast literature and scripts of extraterrestrials and Anunnaki. Water links humans to the Anunnaki.

a-Babylonian account:

In the Babylonian account of the Creation, Tablet 1 illustrates Apsu (Male), representing the primeval fresh water, and Tiamat

(Female), the primeval salt water. These were the parents of the gods. Apsu and Tiamat begat Lahmu (Lakhmu) and Lahamu (Lakhamu) deities. In the Torah, the word water was mentioned in the first day of the creation of the world: "And the spirit of God hovered over the surface of the water."

b-The Chassidut:
In the Chassidut, the higher water is "wet" and "warm", and represents the closeness to Yahweh (God), and it brings happiness to man. The lower water is "cold", and brings unhappiness because it separates us from Yahweh (God), and man feels lonely and abandoned. The Ten Commandments commences with the letter Alef: "Anochi (I) am God your God who has taken you out of the land of Egypt, out of the house of bondage."

c-In Hebrew:
The letter Alef holds the secret of man, his creation and the whole universe (Midrash). In Hebrew, the numeric value of Aleph is 1. And the meaning is:
- **1**-First
- **2**-Adonai
- **3**-Leader
- **4**-Strength
- **5**-Ox
- **6**-Bull
- **7**-Thousand
- **8**-To teach

d-Jewish teaching:
According to Jewish teaching, each Hebrew letter is a spiritual force and power by itself, and comes directly from Yahweh (God). This force contains the raw material for the creation of the world and man. The Word of God ranges from the Aleph (The very first letter) to the Tav (The last letter) in Hebrew.

In Revelation 1:8, Jesus said: "I am Alpha and Omega, the beginning and the ending." In John 1:1-3, as the Word becomes Jesus, the Lord Jesus is also the Aleph and the Tav, as well as the Alpha and the Omega. In Him exists all the forces, and spiritual powers of the creation.

Jesus is also connected to water, an essential substance for the purification of the body and the soul, this is why Christians got to be baptized in water. In Islam, water is primordial and considered as the major force of the creation of the universe.

The Prophet Mohammad said (From the Quran): "Wa Khalaknah Lakoum min al Ma'i, koula chay en hay", meaning: And WE (Allah) have created for you from water everything alive." The Islamic numeric/spiritual value of Aleph and God is 1.

To the Anunnaki and many extraterrestrial civilizations, the An or Alef represents number 1, as well as planet Nibiru, the constellation Orion, the star Aldebaran, and above all the female aspect of the creation symbolized in the Anunnaki's female "Gb'r" (Angel Gabriel to us.)

⌘ ⌘ ⌘

III. Hayya'h, Hawwa and Eve:
Hayyah also means:
- ❖ **a**-Life;
- ❖ **b**-Creation;
- ❖ **c**-Humans;
- ❖ **d**-Earth

In Arabic, Hebrew, Aramaic, Turkish, Syriac, and so many Eastern languages, the Anunnaki words "Hayya'h" and "Hayat" mean the same thing: Life. But the most striking part of our story is that the original name of Eve is not Eve, but "Hawwa" derived directly from Hayya. How do we know this?

Very simple: Eve's name in the Bible is "Hawwa", also "Chevvah".

In the Quran is also "Hawwa", and in all the Semitic and Akkadian texts, Eve is called Hawwa or Hayat, meaning the giver of life; the source of the creation.

Now, if we combine the 2 words: An +Hayya'h or Hayat, we get this: Beginning; The very first; The ultimate; The origin; Water + Life; Creation; Humans; Earth, where the first was created; Woman. And the whole meaning becomes:

The origin of the creation and first thing or person who created the life of humans was a woman (Eve; Hawwa) or water.

Amazingly enough, in Ana'kh, woman and water mean the same thing, because woman as a creative female energy represents water according to the Babylonian, Sumerians and Anunnaki tablets, as clearly written in the Babylonian-Sumerian account of the Creation, Tablet 1.

⌘ ⌘ ⌘

IV. The Anunnaki's equation:

The Anunnaki who created us genetically some 65,000 B.C. lived on earth with us, in Iraq (Sumer, Mesopotamia, Babylon) and Lebanon (Loubnan, Phoenicia, Phinikia).

They taught our ancestors how to write, how to speak, how to play music, how to build temples, how to navigate, as well as geometry, algebra, metallurgy, irrigation, astronomy, you name it. But the human races disappointed them, for the early human beings were cruel, violent, greedy and ungrateful. So, the Anunnaki gave up on us and left earth.

The few remaining Anunnaki living in Iraq and Lebanon were killed by savage military legions from Greece, Turkey and other nations of the region.

The Anunnaki left earth for good. Other extraterrestrial races came to earth, but these celestial visitors were not friendly and considerate like our ancestors the Anunnaki.

The new extraterrestrials had a different plan for humanity, and their agenda included abduction of women and children, animal mutilation, genetic experiments on human beings, creating a new hybrid race, etc...

The Anunnaki did not totally forget us. After all, many of their women were married to humans, and many of our women were married to Anunnaki. Ancient history, the Bible, Sumerian tablets, Akkadian cylinders, Babylonian scriptures, Phoenician inscriptions, and historical accounts from around the globe recorded these events.

You can find them, almost intact, in archeological sites in Iraq and Lebanon, as well as in museums, particularly the British Museum, the Iraqi Museum and the Lebanese Museum.

So, before leaving us, the Anunnaki activated in our cells the infinitesimally invisible multimicroscopic gene of An-Hayya'h. It was implanted in our organism and became a vital composition of our DNA.

Humans are not yet aware of this, as we were not aware of the existence of our DNA for thousands of years. As our medicine, science and technology advance, we will be able one day to discover that miniscule, invisible, undetectable An- Hayya'h, exactly as we have discovered our DNA.

An-Hayya'h cannot be detected yet in our laboratories. It is way beyond our reach and our comprehension.

It is extremely powerful, because it is the very source of our existence.

⌘ ⌘ ⌘

V. The Anunnaki's "Conduit":

Through An-Hayya'h, the Anunnaki remained in touch with us, even though we are not aware of it. It is linked directly to a Conduit and to a Miraya (Monitor, or mirror) on Nibiru. Every single human being on the face of the earth is linked to the outer-world of the Anunnaki through An-Hayya'h. And it is faster than the speed of light. It reaches the Anunnaki through Babs (Star gates).

For now, we will call it molecule or bubble. This molecule travels the universe and reaches the Miraya of the Anunnaki through a Conduit integrated in our genes and our brain's cells by the Anunnaki some 65,000 years ago. But what is a Conduit?

Does every human possess a Conduit?

The answer is yes.

All humans have a Conduit just like the Anunnaki, because it is part of our DNA. It is impossible to explain how a Conduit works inside the human brain, and/or how it works for a human being. The creation of the Conduit is the most important procedure done for each Anunnaki's student on the first day of his or her entrance into a learning center in Ashtari.

A new identity is created for each Anunnaki's student by the development of a new pathway in his or her mind, connecting the student to the rest of the Anunnaki's psyche.

Simultaneously, the cells check with the other copy of the mind and body of the Anunnaki student, to make sure that the Double and the other copy of the mind and body of the student are totally clean.

During this phase, the Anunnaki's student temporarily loses his or her memory, for a very short time. This is how the telepathic faculty is developed, or enhanced in everyone.

It is necessary, since to serve the total community of the Anunnaki, the individual program inside each Anunnaki's student is immediately shared with everybody.

The Anunnaki have two kinds of intelligence:
- **1**-Collective intelligence that belongs to the community.
- **2**-Individual intelligence that belongs to one person. Both intelligences are directly connected to two things:
- **a**-The first is the access to the Community Depot of Knowledge that any Anunnaki can tap in and update and acquire additional knowledge.
- **b**-The second is an individual prevention shield, also referred to as personal privacy.

This means that an Anunnaki can switch on and off his/her direct link to other Anunnaki. By establishing the Screen or Filter an Anunnaki can block others from either communication with him or her, or simply prevent others from reading personal thoughts.

Filter, Screen and Shield are interchangeably used to describe the privacy protection device. In addition, an Anunnaki can program telepathy and set it up on chosen channels, exactly as we turn on our radio set and select the station we wish to listen to.

Telepathy has several frequency, channels and stations. When the establishment of the Conduit is complete, the student leaves the conic cell, where the procedure has taken place, and heads to the classroom.

Now, how does an Anunnaki receive the content of a Conduit to allow him/her to watch over us?

Through the Miraya.

Anunnaki created the Conduit, the Miraya and the An-Hayya'h to watch over us, even though we do not deserve it, said the Ulema. The Anunnaki have been watching us, monitoring our activities, listening to our voices, witnessing our wars, brutality, greed and indifference toward each others for centuries.

But they did not interfere. But now, they will, because they fear two things that could destroy earth and annihilate the human race:

- **1**-The domination of earth and the human race by the Greys;
- **2**-The destruction of human life and planet Earth on the hands of humans.

The whole earth could blow up. Should this happen, the whole solar system could be destroyed. For we know, should anything happen to the moon, the earth will cease to exist.

This is an absolute truth and a fact accepted by all scientists. So anything that could happen to Earth will disrupt the solar system, said the Ulema. An-Hayya'h is our umbilical cord, our birth cord that attaches us to the Anunnaki.

Some refer to it as the "Silver Chord". No matter how silly and crazy this concept might look to many of us, one day, we will accept and possibly we might understand its mind-boggling mysteries, when our science, technology and mind explore wider dimensions, and reach a higher level of cosmic awareness and intelligence, added the Ulema.

Farid Tayarah said: "An-Hayya'h will always be there for you to use before you depart this earth. It will never go away, because it is part of you. Without it you couldn't exist. Just before you die, your brain out of the blue wills activate it for you."

*** *** ***

112. Apindugari: The Anunnaki's Wedding
⌘⌘⌘

I. Introduction:
An Anunnaki word meaning the union (Matrimony) between an Anunnaki male and an Anunnaki female. The union is of a non-physical nature. In terrestrial term it is a sort of a marriage. But the union ceremony is brief, and it occurs through the process of Noura, or the mixing of two lights.

II. Description of the Apingudai: Preparations and the wedding ceremony
In the book "Anunnaki Ultimatum. End of Time: Autobiography and Explosive Revelations of a Human Anunnaki Hybrid", co-authored by Ulema Maximillien de Lafayette and Dr. Ilil Arbel, there is chapter describing the "Mingling of Lights" that occurs during an Anunnaki's wedding ceremony. It is informative and quite revealing.

Herewith excerpts from that chapter:

Note: Ambar Anati is an Anunnaki female hybrid (Half Anunnaki half human). She is the fiancee of Sinhar Marduk, an Anunnaki leader, and soon to be married to him. Here, she is talking to Marduk's sister about the wedding procedures and preparation. Anati has been informed by her mother-in-law, that Anunnaki marriages are quite different from humans' marriages, because a sort of plasmic light is involved in the matrimony process. The

163

marriage is not consumed via a sexual act, but through "Mingling Lights".

The article goes like this:

"He did not exactly explain to me, but I suspect he is indeed apprehensive. He is probably not sure if you know how we are constructed, physically, and what our marriages are really like. I mean, of course you know about the Union, the mingling of lights, but are you aware that Anunnaki do not have sex organs like humans?"

"Of course I am aware of it. I saw a whole lot of little children swimming in the ponds, quite naked, when you took me to our swimming trips.

I know exactly how you are constructed physically."

"Well, does it bother you?"

"Why should it? Miriam, the night before I left, your brother gave me a tiny sample of the mingling of lights. Nothing much, we just held hands and kept apart when he activated our lights, so it involved only the hands. That was all he thought would be appropriate and fair to my free will. So I respected that, but let me tell you, this slight touch was better, a million light years better, than any sexual activity we have on earth."

"So it would not hurt you in any way that as the centuries go by, and you become more and more an Anunnaki, you will lose the sex organs?"

"I would not bat an eyelash."

"You know, Victoria, I believe that your Anunnaki genes influence you considerably already."

"Quite possibly. Also, you must understand that single women in my world usually experiment with sex from their late teens and on. I had a couple of experiences in my twenties."

"Well, I must say I am glad Sinhar Marduchk is too ethical to have spied on you. He would not have enjoyed these experiments, I suspect."

"He might have been made rather happy about them, Miriam. They were nothing to me. Something inside me retained the memory of the mingling of lights with Marduchk, because I really did not like any of these men or what they tried to give me. Sex, as I told an old earth friend just before you brought me here again, is highly overrated...

After experiencing the mingling of the lights, it is a minor thing indeed, a mere physical sensation at best. I don't have to tell you that the mingling of lights involve mind, body, and soul. Just don't tell that to Marduchk... he might feel I needed thirty more years to make sure my free will is intact, and I am not waiting to be sixty before I marry."

"Since I have never experienced sex in the earth way, or ever had sex organs, of course I can't tell the difference. The mingling of lights, to me, is highly spiritual. But do you know, Victoria, that your species is the only humanoid one in the known universes that has sex organs, which were given to you to match the animals and the plants on your planet?

I personally think it was a mistake, it deprived you of true unity with your mates and created much frustration.

No other species share this aberration. Anyway, I am so glad you feel this way, Victoria. I have looked upon you as my sister from the moment you came here the first time, and I would not want to lose you to the earthlings."

⌘ ⌘ ⌘

In the following passage, Ambar Anati is telling us what happened next, and what did she talk about with her husband to be, Sinhar Marduk. Excerpt from the book: "In the morning, Marduchk came early.

I was in the garden, waiting for him, while everyone else was still in their rooms. The morning was beautiful, butterflies fluttered in the dense flower beds and every tree was full of singing birds.

The roses were blooming so intensely that their scent was incredibly strong.

I felt that the occasion meant a lot to both of us, even if I had no idea what was to take place, so I dressed in a semi-formal fashion, and wore a long cream-colored dress with a beautiful red belt.

Marduchk seemed to have had the same thought, since he, too, wore something that seemed more afternoon than morning to me. He was glad to see me up already and asked me to come to the library.

Inside, he approached a shelf, but instead of using one of the usual conical books, he removed an ancient wooden box and brought it to one of the tables. Inside was a scroll that seemed ancient, probably hundreds, if not thousands of years old. "Is this Egyptian papyrus?" I asked.

"Yes, it is. A few thousand years ago, we had to copy the original manuscript. It lasted for eons, but finally it was beginning to crumble. We decided we wanted to make an exact replica, since it was so beautiful, rather than reproduce it on a cone, as many families do."

It was truly beautiful, written in a language I could not understand and illuminated with small, magical pictures, many of them seemingly some kind of symbols. I longed to touch it and see what the texture was, but I did not dare to do so.
"This is our marriage record book where all our marriages are kept. Now is the time to make your final decision, Victoria, if you can. Will you marry me?"
"Of course I will," I said, surprised. "Did you have any doubts about it?"
"I have to ask you formally, no matter what I think. It is part of the ceremony."
"Marduchk, I know nothing about it.
Would you please explain to me what the ceremony is all about?"
"It has four parts. The first is a verbal consent and a statement of free will. The second is signing the marriage record. The third is the mingling of the lights. The fourth is a conclusive ceremony officiated by a representative of the Council."
"And a party?
With beautiful dresses, and music, and food?"
Marduchk laughed.
"Of course," he said. "Right after the ceremony."
"Then let's get on with it, Marduchk!
It's not all solemnity, you know, we should have some fun!"
Marduchk smiled. "Indeed," he said. "We shall now continue, then. Let's start again. Will you marry me?"
"Yes, I will marry you, Marduchk."
"Are you certain, and ready to sign this book?"
"I am certain."
"Can you see, imagine, visualize, or even fantasize any other option than our marriage?"
"No, I cannot."
"And do you realize, that once you sign this book our union is irrevocable?"
"I realize that."

"Is anything other than complete free will ever employed in your consent? Was any form of persuasion, coercion, promises, or any other such behavior employed on my part?"

"There is only free will."

Marduchk unrolled the scroll and picked up a writing implement from the table. It was shaped like a pen, but from previous experience, I knew it worked with energy, not ink. Nothing could erase it, ever. Under a text paragraph, he signed his name in the Anunnaki language.

He then handed me the pen, and I signed my name, in plain English. To my surprise, each of our signatures, though done by the same pen, had a different color. And what's more, under my English signature appeared what seemed to be a translation into Anunnaki, but it kept my signature's color.

"Are we married now?" I asked.

"Yes, we are. For eternity." He seemed suddenly relaxed.

"And then, can I move to your house? Or do I stay at Miriam's house like a nice little virgin until the Council blesses me?"

"No, you move to my house. It is now your house, we are married."

"It's about time," I said, and we both laughed.

Everyone congratulated us when we came back to Miriam's living room. They were all there, waiting for us.

"And now we must take your things to your home," said Miriam, after a good breakfast, where everyone was relaxed, happy, and full of plans.

"Would you believe your husband would not allow me to take your dresses, all the accessories my daughters designed, shoes, and all the jewelry, until after the marriage was accomplished? I knew you will stay, but he just won't allow it, tiresomely refereeing again to free will...but we will try to forget his annoying ways.

We will take everything now, or teleport it, rather, into your closets. Come see all the stuff."

"He does stick to tradition," I said, following her out of the room.

"Ah, well," said Miriam with resignation. "We must accept the fact that all over the known universes, the female of the species is more realistic."

Marduchk heard us, and laughed. "Very true," he said. "Wait one second, Victoria. Once the ceremony and celebrations are all over, where would you like to go for our wedding trip?"

"Why, Paris, of course," I said. "Where else would one go for a honeymoon?"

"What a charming idea," said Miriam. "I was thinking about a planet on the Alpha Centauri system where so many people go for their wedding trips, but Paris will be very nice."

The wardrobe they prepared for me was so incredible, I had to gasp. Closet after closet, full with the most beautiful gowns for all occasions, from casual to formal. Entire shelves full of shoes in all colors, accessories that were everything from hair decorations to belts and scarves, and jewelry that can only be described as breath taking.

"Later, we must decide on the wedding dress," said Miriam. "I have about six or seven for you to choose from."

After I could finally manage to tear myself away from the enchanting wardrobe, everything was quickly teleported without the slightest physical effort, and finally Miriam and Marduchk took me to my new home to get settled.

⌘ ⌘ ⌘

Marduchk's home, which, as I have explained before, was attached with a corridor to his sister's home, was just as beautiful and comfortable as Miriam's. I remembered it from my last visit, since of course I have seen it many times, but it turned out that the brother and sister decided to change everything and make it perfect for a new bride, more elegant, less masculine.

The theme they decided upon was an indoor garden. The living room retained the structure of smooth levels and high ceiling, of course, but it was now filled to the brim with indoor roses, both bushes and trailing, and parts of the floor grew the softest, greenest grass where previously they were covered with carpets.

Other trailing and hanging flowering plants undulated with the soft breeze that came from the opened windows, making the room look like a miniature magical forest.

Tiny ponds, surrounded with ferns and supporting water lilies were on every level, with miniature waterfalls twinkled delicately as they softly fell into the ponds. Tall willow-like trees and bamboo grew behind the furniture.

The couches were all covered with green and gold silk, and the red hibiscus flowers here and there completed the enchanting color scheme. This was the most beautiful room I have ever seen.

The dining room, the guest rooms, the bedrooms all followed the garden theme. Each guest room was decorated with a different flower theme, one of them all lilies of the valley like flowers, another one like an orchard of flowering plums and almonds. The bathrooms were small rainforests.

Our bedroom was Zen like, with the bed in the middle and a few trees behind it and before it. A couple of small tables had each a vase with one white lily, and all the storage for clothes and beddings was hidden behind wooden screens.

One large pond with papyrus growing right in it was placed in a corner, and was surrounded with gray, moss-covered rocks. It was a place of serenity and peace.

As Miriam and Marduchk were conducting the tour of the house, I simply could not stop admiring and exclaiming over everything. It felt to me like something of a childhood's dream or fairy tale. "I am so happy you like it," said Miriam.

"After you live in it for awhile, you may want some changes, some things to make it more comfortable or more suitable for your personal taste. Nothing will be easier. You will tell me, and I'll teach you how to manipulate objects with your mind."

"For the moment, I don't want to even think of changes," I said enthusiastically. "What a place..."

"And now I must go and let you settle. Come for lunch tomorrow, and we'll choose the dress." She left and Marduchk and I were left in our home, finally married and alone in this green paradise. We were now free to bring about our Union and mingle our lights.

Marduchk stood up and said most formally, "Victoria, would you do me the honor of accomplishing our Union? This will be the third part of our marriage."

Tears came into my eyes.

"Do you the honor, Marduchk? No, you are doing me the honor. You, and your family who accepted me as one of them, and the Council who is willing to bless our marriage despite the fact that I am, truly, a member of an inferior species. I am honored." "You are not an inferior, Victoria. You are one of us in every way. You have been brought up on earth, and that changed a few things, but nothing fundamental.

Your genetic makeup is ours, and most important, your spirit and your soul are ours. And the Council will prove it to you by honoring you, during the wedding ceremony, by giving you your very own Anunnaki name."

"The greatest honor is your wanting to marry me, Marduchk, waiting for me, and allowing me the centuries of learning and growing to be like you.

And to answer your question, in the formal style that is needed, yes, I will now happily come with you and accomplish our Union."

There are certain things that cannot be described, things that human language has no words for, and I am afraid that the mingling of the lights is one of them.

I can describe the physical procedure – and unlike human sexual behavior, there is nothing in the Union that is embarrassing or disturbing to even the most traditional and old-fashioned people, but the experience itself is impossible to relate. We went into our bedroom, which traditionally had no windows.

That is because it is not seemly to have the lights, which can be intense, be seen by people who may be walking in the gardens. We sat on the bed, and I said to Marduchk, "I have no idea how the lights are to be activated."

"You don't have to do anything, since the lights are not activated. They emanate from our beings, and just sitting here together will do it. Do you remember how the light grew around our hands, just before you left?"

"Of course I do," I said. I did not want to add that the experience lived in my mind, if unconsciously so, enough to make me come back to him. So I kept it to myself and smiled inwardly with complete happiness.

"Just sit near me, Victoria, close your eyes, and imagine the same light emanating from your entire body." I did as he said, and imagined golden light surrounding me like a soft, flowing veil. For a few minutes nothing happened, and then I felt a change and opened my eyes to see each of us surrounded by a bubble of the most brilliant light.

The bubbles came together, touched each other exactly like floating soap bubbles do, and merged together into one glowing orb.

The light grew stronger and stronger until the whole room was illuminated by undulating, flowing strands of light, a little like the strands that can be seen in the sky during the aurora borealis. And the sensations I felt were the incarnation of beauty, at once mental, physical, and spiritual, since it cannot be anything but a combination of the three, and it mounted and increased until the light exploded into a shower of stars and the Union was achieved. And that is all I can really say, because as I said, human language is too limited. In Anunnaki, there are many words to describe the Union in all its aspects. I do want you to understand, though, I am doing my best.

So if you can recall the most wonderful romantic experience you ever had, with someone for whom you had pure love and respect, perhaps you may have inkling, but only that.

I suspect that until humans evolve mentally, spiritually, and physically, until such time as they can shed all the negative traits of infidelity, jealousy, and fickleness, all they can have is a pale imitation of the Union.

I hope that some day it will happen, because the Union, unlike the human sexual relation, can only ennoble and enrich you, can never be negative, can never cause pain or embarrassment. It is the essence of purity and happiness.

After all was over, and we were resting on our bed, I told Marduchk that I remember smelling the scent of a certain flower I remember from my childhood, which grew only in the hothouses. Marduchk was sure it was part of the experience, and tried to understand which flower it was, but I forgot the name, and could not describe it to him adequately. So Marduchk smiled and said, "Well, make a shower of these flowers fall on us."

"Make a shower of flowers? Me? How do I do that? Don't you forget I am a mere human?"

"Having gone through the Union already started you on developing special powers. Just close your eyes, and imagine the flowers, as you remember them, falling on us like a steady rain from the ceiling. Make heaps and heaps of them."

I laughed at the idea, but to indulge him, I closed my eyes and imagined the flowers doing that. Suddenly I felt something landing on me softly, like a snowflake.

I opened my eyes, and to my total amazement I saw a few flowers falling from the ceiling. White and yellow and smelling like an earthly paradise, the rain of flowers got thicker and thicker and they covered us with their scented petals.

I was speechless with amazement at my new gifts and the impossibility of what I was creating, but Marduchk just picked up one flower and said in a total matter-of-fact way, as if no miracle had been taking place, "Oh, I see, Plumeria. Of course. I

171

should have recognized them from your description; they grow here all over the place." I laughed.

"What next?" I said. "Will I fly to one of the moons on my own silver wings?" Marduchk looked at me seriously and said, "Sure, if you like. There are no limits, really...wings are easy enough to make, any color you want."

What a place, I thought, what a life... and I sank into a blissful sleep under the soft and warm blanket of the delicate while and yellow flowers."

*** *** ***

113. Anunnaki's Extraordinary Powers And Faculties
⌘ ⌘ ⌘

113. Anunnaki's Extraordinary Powers And Faculties
⌘ ⌘ ⌘

I. Introduction:

It would be impossible to list and describe all the Anunnaki's extraordinary powers, in one single tome. In a forthcoming volume, we will add and describe more. In this section, we are going to learn about three spectacular Anunnaki's supernatural faculties.

We have chosen these particular powers, because they are among the most exciting and colorful ones, by our human standards.

⌘ ⌘ ⌘

II. Arawadi:

Arawadi is a term for the supernatural power or faculty that allows initiated ones to halt or send away difficulties, problems

175

and mishaps into another time and another place, thus freeing themselves from worries, anxiety and fear.

A very complex concept that touches metaphysics, esoterism and quantum physics. Ulema Stephanos Lambrakis said that it is very possible to get rid of current problems by "transposing" them into a different time frame. He added that "all of us live in two separate dimensions so close to us. One we know and we call it our physical reality, the other is the adjacent dimension that surrounds our physical world.

Enlightened ones visit that dimension quite frequently. It is a matter of a deep concentration, and perseverance. In fact, it is possible to enter that parallel dimension and leave there all your troubles, and return to your physical world free of worries and problems."

⌘ ⌘ ⌘

III. Araya:
Introduction and definition:
An Anunnaki word meaning prediction code.

According to the Ulema, the Anunnaki's Araya is an effective tool to foresee forthcoming events in the immediate and long term future. The expression or term "foreseeing" is never used in the Ana'kh language and by extraterrestrials because they don't foresee and predict. They just calculate and formulate.

In spatial terms, they don't even measure things and distances, because time and space do not exist as two separate "presences" in their dimensions.

However, on Nibiru, Anunnaki are fully aware of all these variations, and the human concept of time and space, and have the capability of separating time and space, and/or combining them into one single dimension, or one single frame of existence. Anunnaki understand time differently from us, said the Ulema.

For instance, on Nibiru, there are no clocks and no watches. They are useless. Then you might ask: So, how do they measure time? How do they know what time it is now or 10 minutes later, or in one hour from now? The answer is simple: If you don't need time, you don't need to measure it.

However, on Nibiru, Anunnaki experience time and space as we do on earth. And they do measure objects, substances, distances and locations as we do on earth. But they rarely do.

"The Anunnaki (In addition to the Nordics and Lyrans) are the only known extraterrestrials in the universe to look like humans, and in many instances, they share several similarities with the human race..." said Ulema Ghandar.

This physiognomic resemblance explains to a certain degree, the reason for Anunnaki to use time. To calculate and formulate information and to acquire data, Anunnaki consult the Code Screen. Consulting the screen means pragmatically, the reading of events sequences, explained the Ulema. Every single event in the cosmos in any dimension has a code; call it for now Nimera, a "number", added the Ulema.

⌘⌘⌘

IV. Activating the Araya code:
Nothing happens in the universe without a reason. The universe has its own logic that the human mind cannot understand. In many instances, the logic of numbers dictates and creates events. Not all created events are understood by the extraterrestrials. This is why they resort to the Araya Code Screen. Activating the Araya Code requires four actions or procedures:

- **a- Taharim:**
This demands clearing all the previous data stored in the "pockets" of the Net. A net resembles space net as usually used by quantum physics scientists. They do in fact compare space to a net. According to their theories, the net as the landscape of time and space bends under the weight of a ball rotating at a maximum speed. The centrifugal effect produced by the ball alters the shape of the net, and consequently the fabric of space. And by altering space, time changes automatically. As time changes, speed and distances change simultaneously. Same principle applies to stretching and cleaning up the net of the screen containing a multitude of codes of the Anunnaki.

- **b- Location of the Pockets:**
The word "pockets" means the exact dimension and a space an object occupies on the universe's net or landscape. No more than

one object or one substance occupies one single pocket; this is by earth standard and human level of knowledge. In other parallel words, more than one object or one substance can be infused in one single pocket. But this could lead to loss of memory.

Objects and substances have memory too, just like human beings; some are called:
- **1**-Space memory,
- **2**-Time memory,
- **3**-String memory,
- **4**-Astral memory,
- **5**-Bio-organic memory, etc...

The list is endless. Thus, all pockets containing previous data are cleared.

- **c- Feeding the Pockets "Retrieving Data":**

All sorts and sizes of data are retrieved and stored through the Conduit. The Conduit is an electroplasmic substance implanted into the cells of the brain.

- **d- Viewing the data:**

Retrieved data and information are viewed through the Miraya, also called Cosmic Mirror. Some refer to it as Akashic Records. Can the Anunnaki go forward in time and meet with the future? Yes, they can!

An Ulema said that future events have already happened at some level and in some spheres. It is just a matter of a waiting period for the mind to see it.

⌘⌘⌘

V. Ulema Rabbi Mordechai ben Zvi and his student Germain Lumiere "entretien" on Tay Al Ard:

Excerpts from the book "On the Road to Enlightenment: Extraterrestrial Tao of the Anunnaki and Ulema", co-authored by Maximillien de Lafayette and Ilil Arbel:

"Time passed quickly. I have learned so much, everything of which, Rabbi Mordechai promised, would enable me to succeed

in my meeting with the Pères du Triangle, and later in all my endeavors in life. But something,
I knew, was still missing, and I was very hesitant asking about it. One day I gathered all my courage and asked him, "What about the opening of the Conduit?"
"It will happen soon enough," he said.
"But is there work to be done in preparation? What is the process?"
"It varies," said Rabbi Mordechai. "Come on, let's go out, you are tired from so much study." I certainly was, since that evening we continued working after our dinner, having been engaged in an interesting study, so we did not go to our usual walk. It was rather late at night, and I felt I would enjoy a little fresh air. "Let's go to one of the bridges between Buda and Pest," said Rabbi Mordechai. "It's a pleasant night for a walk."
"Which one should we go to?" I asked.
"Let's go to the Széchenyi Lánchíd, the Chain Bridge," said Rabbi Mordechai. I certainly had no objection to that; this bridge was a thing of beauty. It was called after Count Istvan Széchenyi, who had commissioned it, and was the first of the eight permanent bridges in the city.
The Count invested much thought and effort in building the bridge. He not only asked a French authority, Marc Isambard Brunel, for advice, but even went to examine William Tierney Clark's bridge across the Thames at Marlow, England, before finalizing his plans. The bridge was built between 1839 and 1849, and the stunning lions at each end were designed by the great sculptor, János Marschalko.
There is a great debate regarding the lions' tongues. Some say that they are there, though extremely hard to find even if you climb all the way up the pedestals on the four corners of the bridge.
Others say there are no tongues at all, and tell a legend that during the opening ceremony, a little boy noticed that the tongues were not there, and told the sculptor.
Poor Marschalko was so distraught by realizing he had forgotten such an important detail, that he hurled himself off the bridge to his death. It was late at night, there was no one present on the bridge, and the lights of the city reflected beautifully in the dark water.

We stood for a moment, enjoying the sight, and then Rabbi Mordechai said, quite suddenly, "How long do you think it will take you to cross the bridge?"

"I don't know," I said, trying to estimate the length.

"Would you like to bet I can do it quicker than you?" he asked, his blue green eyes twinkling with amusement.

"Sure," I said, laughing. "Why not?"

"Good," he said. "But you must walk straight and not look back or even to the sides." I knew he had something up his sleeve, but it was fun to play the game. "Very well," I said. "Shall I start?"

"Go!" he said, laughing, and I started walking, looking ahead, and avoiding looking back or to the sides. When I reached the end of the bridge, Rabbi Mordechai was standing there, leaning against it, smiling.

"I see," I said. "Very impressive. I would like to learn this technique."

"I am happy to hear that you are not calling it a trick," said Rabbi Mordechai, seriously.

"No, I don't think this is a trick," I said. I felt a vague regret. Have I let Rabbi Mordechai down by being skeptical? Were there some subtle points I have missed?

"Let's go to the other side," he said. "Would you like to try how this technique feels?"

"Yes, I would," I said. In a fraction of a second, I was on the other side of the bridge, without any delay or even any sensation. I was just there, while a second ago, I was elsewhere. Rabbi Mordechai was not near me. I looked at the bridge, and I saw him walking toward me.

Obviously, he wanted to show me that I was not hallucinating. If we were both transported together, I might have suspected that we never really left and it was only some sort of hallucination, another trick, but seeing Rabbi Mordechai walking on the bridge would prevent any such suspicion. He wanted to reassure me, as I thought. I had shown a sad lack of trust, and perhaps I had hurt this great, forgiving, loving friend who would do anything for me. How could I? I felt so ashamed.

When he came to the other side, I said, "Rabbi Mordechai, I know why you transported me and walked yourself. I understand your motive. But it is no longer necessary to do so. I fully trust you. I am your student, forever."

Rabbi Mordechai looked at me with tears in his eyes, and hugged me with all his might. "You are more than my student, Germain. You are my son from now on."

A great wave of happiness and peace flowed through me. He was not angry, he understood, he knew I placed all my trust in him and I was forgiven...Suddenly, I felt something I could not explain, something that happened in my mind, or in my brain, or in my soul, something that I could not prove but was as tangible as the river and the houses.

The ability to trust I have so suddenly discovered in my self burst open the gates in my mind, and my Conduit opened.

I staggered a little, caught on to the bridge, and recovered almost instantly. The world felt different than before, but I was still myself.

"How did it happen so fast?" I asked.

"It was not fast at all," said Rabbi Mordechai. "It was exactly as it should be, as it always is, and always will be. You see, your other masters taught you many things, and there was an enormous amount of dormant knowledge which was accumulated in your mind and constantly fed by them. And now, at the right time, and under the right circumstances, and encouraged by your ability to accept the Ulema way, the Conduit opened, by itself, like a flower. You are now ready to start on the road to being full-fledged Ulema. Welcome, my son."

This story was told by Germain Lumiere, and written by Ulema Maximillien de Lafayette

*** *** ***

114. Alien Mirrors That Remember Their Previous Owners
⌘⌘⌘

183

114. Alien Mirrors That Remember Their Previous Owners
⌘ ⌘ ⌘

I. Introduction:

This is one of the most striking and mind-twisting stories in the annals of the occult, extraterrestrial studies, parapsychology, and alien phenomena.

It is part parapsychological, biological, extraterrestrial, macabre, and even scientific.

The subject was discussed by members of prestigious French academies of science, noted historians, and most recent a learned group of Anunnaki-Ulema.

II. The story:

This bizarre event took place in France in 1997. It would be fine about 500 years ago, during the struggle with witches, but it happened at the end of the 20th century, the most rational century. The event was extraordinary: Antique dealers addressed journalists with the request to warn the collectors of antique things not to buy the mirror, which had the inscription Louis Arpo 1743 on the frame.

They said that the mirror had killed almost 38 people during the long history of its existence.

Antique dealers decided to address the media as the mirror had disappeared. It was found missing when a criminal law professor asked to take photographs of the mirror to show them at his lectures.

"The mirror was kept at a police station after it had killed two people in 1910. However, someone penetrated in storage facilities and stole several things, including the mirror. We think that the thief will try sell it, that is why we are trying to provide as much information as possible about the mirror to warn potential customers of the danger," a spokesman for the French association of antique collectors said.

The mentioned mirror provoked a cerebral hemorrhage for the people that were looking into it. Some people believed that the mirror was reflecting rays of light in a specific way, whereas other believed that it happened because of the mirror's negative energy that it had saved for hundreds of years.

There were some people, who thought that the mirror was a window to the other world. There is no common opinion about the antique mirror, but people still try to explain the reason of mysterious deaths that it had caused.

III. Tommaso Campanella's explanation:

A mirror is like a magnet, it is capable of attracting poisonous evaporation and accumulate it on its surface. Italian philosopher Tommaso Campanella (1568-1639) had a rather scary theory about it.

He stated: "Old women, who do not have warm blood in their veins, who have horrible smell coming out of their mouth and eyes, find that their mirrors become blurred, because moisture drops of their heavy breath stick to the clear cold glass. If their saliva drops on fabric, the fabric decays.

If a child sleeps with an old woman, its days will become shorter and her days will become longer, and a child will die." (Source: Tommasso Campanella, Del senso delle cose e della magia, IV, 14.)

IV. Paris Academy of Sciences:

French scientists discussed the same question in a hundred years too. A document of the Paris Academy of Sciences, dated 1739, runs: "When an old woman approaches a clear mirror and spends a lot of time in front of it, a mirror absorbs a lot of her bad juices. The chemical analysis showed, the juices were very poisonous."

Some researchers use this property of mirrors to explain a superstition, which says that one should not come up to a mirror during an ailment. Poisonous substances that a sick person breathes out stay on the glass of a mirror and then evaporate, causing damage to the health of the people, who inhale the poisonous air.

V. Parapsychology:

Researchers stated that it was not the chemical sediment of the Louis Arpo mirror that had killed its owners.

Poisonous substances on a mirror can be washed away with water very easily.

The mirror has been certainly washed many times. However, a mirror might save some information - it would not be insane to suppose that a mirror has a memory.

A mirror in a house witnesses all events that happen there. A mirror reflects dramas and tragedies, stupid and funny things, it reflects beauty and ugliness. At times, people wish a mirror could show episodes that it used to reflect years ago, to playback the past.

A. Vulis wrote in his book *Literary Mirrors*: "A mirror is always the present, without the past and without the future. A mirror is the incarnation of amnesia. It is a streaming moment that disappears forever once it has been reflected."

It is really very hard to believe that a mirror is like a video camera, but it is not excluded that mirrors are capable of remembering something.

Visual images are not likely, but it is possible that a mirror can "remember" peculiar features of its owner, like other things can.

VI. Psychology and metalogic:

Doctor Hans Berendt from Israel conducted an interesting experiment with a woman, who had extraordinary sensual abilities.

He asked her to explain the feelings that she experienced from unknown objects inside two identical boxes. The woman said that she had felt a strange feeling of a push from one box, but the thing in the other box was completely different.

She said, there was something ancient about it, a dilapidated amphitheatre with huge ancient amphoras. When the doctor opened the boxes, the woman found windowpane pieces in one of them and ancient Roman coins that archaeologists had found in Bethlehem.

Glass fragments in the first box were taken from a window that was broken with a powerful blast in Jerusalem that had killed dozens of people.

If all things have certain memory properties, a mirror is not an exception, especially when it comes to the mirrors with silvery amalgams, because silver is the metal of a strong informational capacity.

One should assume that a mirror can radiate the information that it has saved before. This radiation might affect a human being. One should be very careful before hanging an old mirror in a room. It might be filled with the negative energy.

An old mirror with rich history often causes strange dreams full of bizarre images; it might evoke unusual desires, inexplicable fears and so on.

Most likely, a mirror remembers the condition, emotions and feelings that a human being had.

Most likely, a mirror keeps the information, and remembers not images, but the meaning of them. This meaning might affect the mind of a new owner.

VII. Extraterrestrial influence:

An Anunnaki-Ulema stated that "everything in the universe has a memory, including matter. Thus we have spatial memory, metal memory (memory metal), time memory, emotional memory, parallel memory, multidimensional memory, so on.

All objects retain a depot of events and vibes that have occurred around them.

Some of the depot's contents is of a positive nature, while other parts of the contents are of a negative nature. If an object is of an extraterrestrial origin, it could radiate fatal vibes that affect the mind. Health will be deteriorated, and death becomes eminent."

One of the mirrors reported in the Anunnaki-Ulema literature is called Miraya, a very sophisticated tool allegedly used by the Anunnaki to monitor time and space, and activities of other extraterrestrial races.

The Miraya can be deadly if used by humans.

Another type of mirrors is called Mnaizar, a sort of table-magnifier made out of plasma-energy, Ulema use to track events and produce supernatural phenomena such as teleportation, and metals' transmutation.

The Mnaizar is a highly sophisticated instrument originally developed by the Anunnaki while they inhabited Earth.

The Ulema have several copies of these mirrors. Ulema W. Li stated that it is very possible that the French mirror is some sort of, or a version of the Mnaizar.

It retains memory, and even project holographic calendar of future events. Humans who gaze at this sort of mirrors will expose themselves to a great danger.

The Mnaizar emits undetected fatal radiations."

*** *** ***

VIII. Jazirat Arwad And The Anunnaki's Connection
⌘ ⌘ ⌘

Arwad "Aradus"
I. Definition and introduction
II. The secrets and unknown historical facts of Arwad

I. Definition and introduction:

Phoenician/Ugaritic/Greek noun.

Name of an ancient Phoenician island in the Mediterranean Sea.

The island of Arwad was an independent kingdom in the days of the Canaanites.

It was created by the Phoenicians in early second millennium B.C.

This small beautiful island located 5 miles from the city of Tartus in Syria, was one of the first Anunnaki's small colonies on Earth.

It was mentioned in the Bible by the Prophet Ezekiel. Arwad was the headquarters of the seven wise men who came from Apsu, the sweet water, and attended the gods of Enki.

They were known to the Sumerians as Abgal, to the Akkadians as Akkallu, and to the Phoenicians as An-Khal.

The Anunnaki called them "The Ab-n'GAL."

On the island of Arwad, the Phoenician created a secret society called the "Circle of the Serpent" to honor their god Melkart.

On Arward, the Melkart shrine/altar still stands in all its beauty and majestic architecture.

The early learned Greeks who visited Arwad studied medicine at the Phoenician-Anunnaki medical center, and when they returned home, they adopted the Phoenician sign of the serpent as the logo for their healing arts.

II. The secrets and unknown historical facts of Arwad:

Arwad hold many secrets and unknown historical facts, to name a few:

191

- **1**-For a short time, Jesus and Mary Magdalene lived there after the Biblical Crucifixion.
- **2**-St. Paul sailed to Arwad after he has spent some time in Byblos (Jbeil) and Batroon in Phoenicia (Today, Lebanon).
- **3**-It was at Arwad, that the Anunnaki created the "Brotherhood of the Serpent".
- **4**-The Phoenicians had a secret society called "The Fish" and it was headquartered in Arwad.
- **5**-The Templar Order used the island as a hide-out. In fact, Arwad sheltered the last Crusaders and the remnants of the Templar knights who fled France following their massacre on the hand of the king of France and the infamous Inquisition.

Some claimed that the Templar knights returned to Arwad to retrieve the Holy Grail; the genealogical tree of Jesus and a buried gospel by Mary Magdalene. Ironically, at one time in history, Arwad was Pope Clement V's gift to the Templar knights. The Island of Arwad was the last stronghold of the Crusaders in the Near East.

- **6**-During the French occupation of Syria and Lebanon during the Second World War, the Vichy French Government discovered Anunnaki-Phoenicians tablets buried underground in Arwad.

The ancient tablets told the story of a race of super humans who descended on earth and taught the fishermen how to navigate the sea and how to read the maps of the stars. They created a secret society called "Brotherhood of the Fish".

Later on in history, the "fish" became the secret symbol of early Christians. It was St. Paul who first created the fish symbol as a secret way for early Christians to recognize other Christians in the Levant, Greece and Rome. In ancient times, Arwad was a refuge to many persecuted Phoenicians, Hebrews, as well as Greeks and Romans. It is a perfect spot for those who are fond of Ernest Hemingway. You will be intrigued by the layout of the island's houses and its fortress.

*** *** ***

116. Baalbeck, The Annunaki's City on Earth
⌘ ⌘ ⌘

I. Definition and introduction
II. Baalbeck, the Anunnaki, the Phoenicians, and the Sumerian/Akkadian texts
III. Baalbeck: A visit to the underground city of the Djinn and Afrit

---

I. Definition and introduction:
Name of a legendary city in Lebanon. The site of Baalbeck dates back 4000 years, B.C., when the Canaanites built a temple to worship "Baal" the Semitic God of thunders and storms.

In the Hellenistic ages, Baalbeck was called Heliopolis "The City Of Sun" as it is known till nowadays, identified with the Greek Sun God "Helios". Baalbeck is one of Lebanon's oldest cities, and one of the most important Roman sites in the Middle East.

According to his Eminence Estfan Doweihi, the Maronite Patriarch of Lebanon: "Tradition states that the fortress of Baalbeck is the most ancient building in the world. Cain, the son of Adam, built it in the year 133 of the creation, during a fit of raving madness. He gave it the name of his son Enoch and peopled it with giants who were punished for their iniquities by the flood."

II. Baalbeck, the Anunnaki, the Phoenicians, and the Sumerian/Akkadian texts:
Baalbeck (Baalaback in Arabic and Lebanese) is one of the oldest habitats on earth, built by the early Phoenicians and the Anunnaki during their first and second landings on earth. Baalbeck was the original space mission center of the Anunnaki.

Today, a launching pad for extraterrestrial spaceships is still visible at Baalbeck nearby the Temple of Jupiter.

Later on in history, Baalbeck became a major occult and a healing center visited by many kings and emperors. Attracted by its beauty and supernatural properties, the Roman emperor Augustus made Baalbeck a Roman colony and a major oracles shrine.

In fact, the Roman emperor Trajan consulted a celebrated oracle in Baalbeck. Unfortunately, Baalbeck was totally sacked and decimated by the Muslim Arabs in 748 A.D.

In 1,400 A.D., the Muslim Turkish conqueror Tamerlane pillaged and destroyed the city, and several Roman-Phoenician-Anunnaki temples and altars. In 1,759 A.D., a major earthquake decimated the remaining ruins and almost all what was left from the Anunnaki-Phoenician monuments.

There is one place on earth, the Ulema consider as the ultimate "terminal" of the Anunnaki; a sort of a Ba'ab from which a person enters or exits a physical dimension. And that place is Baalbeck.

Thousands of years ago, and long before the Sumerians established their kingdom in Iraq, and interacted with the Anunnaki, and many many centuries before the human race in any region of the world learned about God or Gods, the Anunnaki landed in Baalbeck, and revealed to its inhabitants many secrets, including teleportation, psychic healings, and the divine nature of the supreme beings (Gods, creators).

Baalbeck served them as a landing and a launching post. It still exists today. Baalbeck was mentioned in ancient Sumerian, Babylonian, Assyrian, Akkadian, Chaldean, Hittite, and Persians epics, texts, and tablets.

Baalbeck was the rendezvous and favorite sacred place of the kings and deities of Sumer, Babylon and Egypt, because it was the first city established by the Anunnaki on earth.

The legendary king of Uruk, the Anunnaki king-god Gilgamesh visited Baalbeck, and worshiped there. He worshiped higher gods. Why did he go to Baalbeck?

What did he expect to find?

Gilgamesh hoped to acquire immortality and extra supernatural powers from the gods who lived in Baalbeck. The gods welcomed Gilgamesh and told him Baalbeck is the entrance to the other world; to the primordial sphere that created Earth and the human race.

The gods of Baalbeck directed Gilgamesh to the secret celestial Ba'ab of his ancestors the Anunnaki; from that Ba'ab (Exit), Gilgamesh could reach Ashtari in a blink of an eye.

The Gods also told Gilgamesh that as soon as he enters the Ba'ab he will become immortal, but he should continue his journey to Ashtari (Nibiru), and Gilgamesh did. And Gilgamesh asked the gods: "Will the Ba'ab take me directly to Ashtari, so I would reach immortality?" and the gods answered: "Eventually, but first, you must make a short stop at the spring of immortality at Al Arz, a sacred region of your ancestors, the Anunnaki..."
And Gilgamesh asked again: "Where is Al Arz?"
And the gods answered:

"Not very far from here...
it is the highest and mightiest mountain in Phoenicia,
where the Anunnaki your ancestors
planted the cedar trees...
and on the top of the mountains
you will land for a short time
where you will clean your thoughts and mind...
nd you will stroll under the branches
of the cedar trees...
and short after you will continue your journey
to Ashtari...and immortality you shall have..."

Al Arz is a mountainous region in Lebanon (Ancient Phoenicia), where the Biblical cedar trees grow; the same trees King Hiram of Tyre used to build the Temple of Solomon, and King Tut An'k Amoon used to decorate his palaces. Also Noah's boat was built with Phoenicia's Arz trees.
Al Arz means cedar trees in Arabic and Phoenician.
It is derived from the Anunnaki's language "An'k".
The Christian Lebanese revere Al Arz in their Catholic Maronite mass and implore the Cedar trees for favors and heavenly support in the same way they pay reverence to Virgin Mary. In their Maronite Catholic mass, they say: "Ya Arzet Loubnan Salli Li Ag Linah" meaning: "O Cedar of Lebanon pray for us."
Just lovely!
They strongly believe that the cedar trees are immortal.
The Sumerians believed that Al Arz and Baalbeck are the holy cities where immortality lives as disguised gods on planet earth.
This is why Gilgamesh traveled to these two old Anunnaki-Phoenician cities.
The best-known and most popular hero in the mythology of the ancient Near East, Sumerian texts and Anunnaki's literature, was

Gilgamesh, a Sumerian king who wished to become immortal. Gilgamesh was the son of the goddess Ninsun and of either Lugalbanda, a king of Uruk, or of a high priest of the district of Kullab.

Gilgamesh's greatest accomplishment as king/god was the construction of massive city walls around Uruk, an achievement mentioned in both myths and historical texts.

Gilgamesh first appeared in five short poems written in the Sumerian language sometime between 2000 and 1500 B.C.

The poems "Gilgamesh and Huwawa", "Gilgamesh and the Bull of Heaven", "Gilgamesh and Agga of Kish", "Gilgamesh, Enkidu, and the underworld," and "The Death of Gilgamesh", relate various incidents and adventures in his life, and his obsession with immortality.

III. Baalbeck: A visit to the underground city of the Djinn and Afrit.

The following is an excerpt from the book "On the Road to Ultimate Knowledge: Extraterrestrial Tao of the Anunnaki and Ulema", co-authored by M. de Lafayette, and I. Arbel.

Chapter Three: Germain Lumiere, as a young Anunnaki-Ulema is telling the story of his visit to Baalback.

I graduated from high school at seventeen, and was free for a while. Much thought had to be given to the decision and preparations for my university studies. I expressed a desire to go to Paris for that purpose, and was considering various fields, but nothing was final. I did not mind a little time off, though, and enjoyed my temporary idleness very much. At that time, the Master was visiting us, and as usual, had an incredibly exciting plan for me.

"Have you ever been to Baalbeck?" he asked.

"No, never."

"It's an interesting city, very old. There is a lot of controversy as to who built it, though."

"Isn't there some historical evidence?"

"Plenty, but there are four interpretations. The Christian Lebanese say it was built by the Phoenicians. The Muslim Lebanese prefer a theory claiming it was built by Djinn and Afrit. Some important occult leaders say it was built by Adam, after he was kicked out of Paradise. Well…"

"And the Ulema, what do they think?" I asked, knowing that this was the theory I would trust.

"The Ulema say it was built by the Anunnaki and the proto-Phoenicians who lived on the island of Arwad and in Tyre. There is a lot of evidence in this direction."

"So will I see the ancient parts?"

"Of course. I would like to take you to a very special part of the city, where the Founding Fathers of the Ulema used to meet thousands of years ago. Unfortunately, we no longer meet there, because it became a tourist attraction and a state-controlled center of music and dance festivals. It will be fun for you, though, to mingle with all these tourists, it's a nice place."

"But surely that is not the reason for going," I said.

"No, it is not. I plan to take you to a secret underground city under Baalbeck, and show you where the Anunnaki landed for the first time on earth. Very few people know what is going on under the modern city of Baalbeck.

The first Anunnaki landing took place before the Deluge, though they came again and again after the Deluge as well."

"Before the Deluge? When was that, exactly?" I asked.

"About 450,000 years ago, perhaps a bit longer. At that time, the Anunnaki created the humans."

"And what about God?" I asked. Even though I was taught much of the Ulema traditions and world view, I never heard about the creation of the human race.

"No one ever heard of God 450,000 years ago. You start to hear about God only around 6,000 years ago," said the Master. I knew enough about the Anunnaki at that time to accept this without much trouble, so I went to find Mama and Sylvie and tell them about the upcoming trip.

The trip from Damascus to Baalbeck could be accomplished in about two hours, at least you could do that if you traveled in a decent car. We took a bit longer to get there, since the car, borrowed from a friend of the Master who was also to drive us there, was an ancient Mercedes that did not use normal gasoline but rather employed *mazut*, or diesel fuel, and made such a racket it was impossible to hear yourself think. To my surprise, I saw a mysterious Sudanese man sitting in the back seat, dressed in ill matching jacket and pants and scowling at us.

At the Master's request, he started to get out of the car to introduce himself. I watched the process in fascination, since he was not doing it quickly like a normal person, but instead was slowly extricating himself in stages, gradually disentangling himself, like a huge snake. I have never seen such a tall man, or anyone as strange. He was about seven feet tall, very thin, and his face did not look quite human to me, but like a giant from outer space.

This bizarre apparition just stood there, looked fierce, and played with a string of amber beads. The Master ignored his uncouth behavior and introduced us.

"This is Taj," he said. "His name means 'Crown.' He is joining us because he has the key to the gate of the secret city underground. He is also able to persuade the Djinn and the Afrit to open certain doors, which is quite a talent." I was not sure if the Master was joking about the Djinn and the Afrit, so I kept quiet, nodded to the Sudanese, and got in the back seat.

Taj folded himself back into the car and sat beside me, the Master went into the front seat, and the driver, who seemed to be normal and cheerful, greeted the Master and me in a friendly way.

The car started making a noise that was worthy of demons, but I did not care because I was thinking about the real devils, the Djinn and the Afrit.

I leaned forward and asked the Master, "Would I be able to see the Djinn and Afrit?"

"Yes, of course," said the Master casually. "You can even try to talk to them, if you like. The underground city is actually called the City of the Djinn and the Afrit; plenty of devils are there." Since these devils did not seem to frighten the Master, I assumed he knew what he was doing, and sat thinking about what my part could be in this unbelievable adventure. However, I was aware of increasing irritation by what Taj was doing. He constantly played with his amber beads, clicking away on and on. I asked, "Why do you have to click these things all the time?"

Taj seemed annoyed by my question. "Try them yourself," he said curtly, and handed them to me.

I grabbed at them, and instantly, a horrible electric shock went through my entire body, quite painfully, and I cried out and threw the beads on the floor of the car. The Master screamed at Taj, "How dare you? How many times did I tell you never to do that? Give me the beads immediately!"

Taj handed him the beads, meekly enough, and had the grace to look embarrassed. The Master rubbed the beads, seemingly absorbing and removing the energy, and then returned them to me. "You can try them now," he said. "And don't give them back to Taj until I tell you to." Taj said nothing. He seemed unhappy in the car, constantly fidgeting, and could not sit still. Perhaps he was claustrophobic, I thought, and the confined space bothered him. We drove on.

Finally we arrived in Baalbeck. "Where now?" said the driver.

"We are going to the *Athar*, the ruins," said the Master.

"I don't know how to get there," said the driver. "Shall I ask for directions?" He parked the car. There were many people around, some Arabs in traditional garb, some Europeans in every kind of attire and carrying backpacks and cameras. It seemed to be such a normal, cheerful place. I thought of the festivals and the music; how could there be Afrit and Djinn and all sorts of underground labyrinths in a place like that? It was as modern as can be.

"When you are with Taj, you do not ask for directions," said the Sudanese with a superior air. The driver shrugged, not quite convinced.

Taj winked at me and stared at the driver's neck, concentrating. The driver suddenly started to beat his own neck, complaining how much he hated mosquitoes. I was certain there were no mosquitoes in the car, and I was sure that Taj created the imaginary insects that were tormenting the driver.

The driver's neck became really red.

"Taj, stop this nonsense immediately!" said the Master severely. Apparently, Taj could send certain energy rays that had the capacity of annoying people. Taj stopped, gave the driver the necessary directions, and we went to the Athar.

"First, let's go to the world biggest stone," said Taj. We drove further, and as we turned a road toward the Temple of Jupiter, I was shocked by the sight that met my eyes. It was a huge gray slab, partially buried in the sand, perfectly cut and smooth. It was unquestionably man made, not a natural formation, a short distance from the Temple. How in the world could such a stone get there? Who could have carried it? This stone was so immense that the stones of the Egyptian pyramids would be infinitely small, completely dwarfed, if put next to it. The Stonehenge monoliths would be insignificant if they were placed next to it. In addition, it was immensely old, and even modern equipment could hardly cope with such a giant, let alone ancient technology.

"How big is this stone?" I asked, truly awed by the sight.

"Seventeen hundred tons," said the Master.

"It is hand made, isn't it?" I said. "It is too straight to be natural. It simply can't be natural. And yet, how could it get here, if it is artificial? It just can't!"

Taj grinned and said, "Hand made, yes, but not by human hands."

I was beginning to get the idea. "Then who made it?" I asked.

"It was part of the landing area used by the Anunnaki," said the Master. "There are six stones like it. Only the Anunnaki could move such a slab."

"Ah, but I can make it fly," boasted Taj.

"You must be crazy," I said, disgusted with him.

"You want to see?" He said.

"Sure," I said. "I would like to see you do that."

"Very well, but not when so many people are around. We will be back around nine o'clock, no one is around, I will show you."

Since it was around four o'clock in the afternoon, I was wondering how we would spend the time, but the Master had his own plan.

"We have plenty of time to do what needs to be done," he said. "I would like you to meet Cheik Al Huseini." This was the first time I met the great man, who later contributed greatly to my studies.

We went back into the car, and drove to the Cheik's house. The house was small and modest, built sturdily of stone, with thick walls. The door was low, as was normal for middle class Arab houses. This style was followed for many years, for the sake of safety and security.

Apparently, the conquering Ottomans used to sweep into houses that had large entrances while riding on their horses, and thus be able to kill and destroy anyone and everything inside. The low entrances forced the rider to get off his horse first, making him much less dangerous to the inhabitants.

In the big living room, which they called the *Dar*, many sofas were placed against the walls, arranged next to each other.

About twenty to thirty men were present, dressed in Arab robes and turbans. All were elderly, with long white beards. The Cheik was sitting in the place of honor. When the Master arrived, everyone stood up, repeating the word "*oustaz, oustaz*," to each other, meaning "teacher." Someone pointed at Taj, and said, "The Afrit is already here." I thought this description fitted Taj perfectly, but expected him to be angry. To my surprise, he seemed pleased by being called that name, and grinned at me like a delighted child.

We sat down, and the men came to kiss the hand of the Master. The light was low, only one lamp was turned on, but I could see that one person did not get up from his seat. Since this was strange behavior, I looked at him carefully, and to my amazement recognized the old Tuareg, whom I had met years ago in the suk in Damascus, the man who was cut in half. He recognized me as well, smiled, and motioned to me to come and sit by him. I came, and he said jokingly, "Don't start searching for the rest of my body..."

I laughed, a little guilty, because that was exactly what I was planning to do. At any rate I could see nothing, since the long robe he wore covered everything. Everyone conversed in Arabic, which by now I spoke very well, and after a while the Cheik motioned most people out. Eight of us remained in the room. The Master, Taj, and myself, were the only outsiders. The Cheik, the Tuareg, and three other elderly Arabs completed the number of the people who were permitted to attend.

At that moment, a man came from an inside room, carrying a big copper pot, full of steaming hot water. He put the pot on a table in front of the Cheik, addressing him by the title *Mawlana*.

This title meant "you are a ruler over me," and was used only to address kings, sultans, or prophets. I was surprised. This title belonged to very important people, but the house and everything in it spoke of middle class. So what could this mean? The Cheik must have been a very important person, somehow. I planned to ask the Master about it later, not wishing to disturb him with questions at the moment, since I was sure strange things were about to begin to occur.

I was sitting near enough to the Cheik to see everything very clearly, and waited breathlessly for the events that were to come. The Cheik took three pieces of blank paper, and threw them into the hot water in the copper container.

The room was completely silent, no one moved, except Taj, who whispered to me, "You are going to like what you see, it's fun, but don't move no matter what happens." I nodded, and concentrated on the pot, looking occasionally at Taj for clarification.

Somehow he assumed the role of my guide to the occult world, and I realized he knew exactly what was taking place. "Shush, just look at the container, something is about to happen," he said. I went on staring at the pot.

Suddenly, in a blink of an eye, the water in the container disappeared, and the three pieces of paper burst out of the container. They lined up in the air, without any support, one after the other. They waved about for a few seconds, then merged and became one larger piece.

The piece of paper started swirling in the air, rotating around itself, quicker and quicker, and suddenly stopped in mid motion. It was suspended in the air, completely still, and in a flash, letters appeared on it, printed clear, black, and easily visible from where I was sitting, though I could not make out the words.

The Cheik got up, approached the paper, read the words, and then asked one of the people attending to close the shutters on all the windows. The room became very dark, and the words, seemingly separated from the paper, glowed in air like a bright hologram.

The Cheik called Taj, and asked him to read the words. I could not hear what they said to each other, but they seemed to agree on something, as they stood there, nodding their heads. Then Taj came back to me.

I asked him, "What was that?" He stepped on my foot to quiet me. His large foot's imprint was painful, so I shut up. Everyone else seemed to accept the phenomenon without trouble, and gazed at the Cheik as he began to move in a strange manner.

He looked to the left, mumbling something incoherent, then to the right, saying the same incomprehensible things, repeating the sequence twice.

Then he lifted his hands as if in prayer, in the manner shared by both Jews and Muslims. Touching his chest and pushing his hands in front of him, he said, *"Ahlan, ahalan, ahlan, ahalan, bee salamah."*

The letters were still glowing in front of him in the air, and he added, *"Asma' oo hoosmah ath sab'ha."*

I turned and pinched Taj, whispering feverishly, "Explain!"

"Don't you know anything?" said Taj. "These are the names of seven Afrits. They are going to open the gate of the underworld for us."

"But..." He stomped on my foot again to shut me up, and it really hurt and I kept quiet.

The Cheik said, rather loudly: *"Bakhooor, bakhooor!"* A man appeared out of nowhere and brought an incense holder. The Cheik moved it back and forth, the room filled with smoke, and everyone started to chant and mumble very loudly.

I understood nothing at all of what they said. It seemed they were speaking in tongues, and the effect was frightening. They went on for a couple of minutes, then stopped abruptly. At that instant, the letters pulled together, became one shining ball of light of intense silver color, and zoomed out of the room into thin air.

One of the people opened the shutters and the late afternoon light streamed in. The Cheik put his right hand on his heart and said "Thank you" three times.

I was wondering who exactly he was thanking, and who, originally, was he praying for, since he never used the words God, Allah, or any other recognizable deity name. I did not realize at the time that the Ulema, even when they were Arabs, where not Muslims, and had their own, very different, world view.

The Master got up. Everyone rose with him, their robes swishing and making a faint sound in the quiet room. The Tuareg floated in the air. I looked at him, doing my best to control my discomfort. His upper body was solid, but the bottom half of the robe was obviously empty as it swirled around him, making the absence of his lower body extremely and disturbingly clear. He seemed like an apparition, a ghost.

Everyone came to the Master, bowed to him, and then grabbed his hand with both of theirs, in a way that was clearly ceremonial. The right hand's thumb was hitting the spot between the thumb and first finger of the left hand, and then the left hand covered the right hand.

The Tuareg floated near the master and did the same thing. Everyone looked at each other and thanked each other a few times, following their thanks with the words "*Rama Ahaab*." I did not know this word, and was not aware that they were speaking Ana'kh, the language that was shared by the Anunnaki and the Ulema.

And yet I sensed that there was something very special about the way they spoke, as if by instinct. I was staring at the people and trying to understand their words until the Master tapped me on the shoulder and told me to come out.

Taj left with me, and said, "You talk too much. You should be paying more attention, such an occasion is not likely to happen again!" I shrugged, but I had to admit to myself that he was right, these events were probably unique. To my surprise, I was beginning to like the Sudanese, and no longer felt threatened by his strange appearance and bizarre behavior.

As if reading my mind, he put his hand in the inner pocket of his ill fitting and flashy jacket, pulled out two lollipops, and handed me one.

"Won't you tell me a bit about the Afrit?" I asked, licking my lollipop. "I am not sure why we need to call them. Why can't we just go into the underground city? I don't quite understand anything that is going on here."

"In your home, in France, do you have a *Jaras*, a bell, on your door?" he asked.

"Yes, of course," I said, surprised at the question.

"Well, you see, the underground city do not have a Jaras, and it is locked. If you want to come in, someone must let you in. The Afrit can help you, but you have to call them in a special way. Otherwise, they don't know you want them to open the door. How would they know? They are not too clever."

"Where is the door?" I said.

He pointed to the ground. "Under you, under the house, there is a door. Right under the Cheik's house. A door to the *Aboo*, the deep abyss. It is also called *Dahleeth*, meaning an underground labyrinth."

"Are there other doors?"

"Very likely, but I only know this one."

One of the people came out of the house, motioned to us to come in, and said, "We are ready." In the house, everyone was wearing a white robe, and to my surprise, their heads were covered with the type of head scarf Jews sometimes wore in the synagogue. To confuse the issue even further, one was holding a scroll that resembled a Torah.

I felt desperate. Were they going to delay our journey again and start praying? I really wanted move on, see the Afrit, have the adventure. I was tired of the delays. Thankfully, one of them handed me a robe and commanded me to go change my clothes, which I did, but Taj did not change his attire.

I asked him why he was not required to do so, and he explained that he was not one of the *Al Moomawariin*, or the enlightened ones, so he was not required to wear the special outfit.

This did not really clarify the matter, since I was not one of the enlightened ones either, but I decided to let it pass.

Taj seemed to be right about the door being under the Cheik's house, because we started to descend the steps to the basement. The basement was long and narrow, and had a very high ceiling, perhaps the height of two stories. Everything, floor, walls, ceiling, were made entirely of gray cement. It smelled of dampness, and was very cold.

We went through a one room after another, all narrow and long, eventually reaching a small room that had an iron gate by its far wall. The Cheik opened the gate with a large key, and behind it was a second door, made of thick wood. A second key opened this one. Suddenly a thought struck me.

Why did he need a key?

Why couldn't a man who had such supernatural powers simply command the doors to open? Or pass through them like a ghost, for that matter?

I expressed my thought to Taj. "It won't work," said Taj. "Yes, of course the Cheik could pass through doors, but how would he take you with him?"

"What do you mean?" I asked, bewildered.

"You are not enlightened as yet. You cannot use supernatural means of transportation at this stage, so if he wants you, or me, for that matter, to pass through these doors, he must take you inside in a normal way. If he tried, you will just bang against the doors and hurt yourself, while he would be on the other side." I began to see that Taj was not stupid at all. Childish, and sometimes pretending to be silly and play silly games, but deep down, he was extremely knowledgeable.

We stood together in the small room, exactly like all the other rooms in the basement. The Cheik said, "Let the boy be the last one. He needs protection. Taj, come here."

Taj joined him at the front of the line, and we entered a long corridor. As we were walking, the corridor began to shift its shape.

I felt seasick, nauseated, my balance was lost. The floor, and walls, everything was moving, rolling, undulating.

I did not see clearly, and wondered how long this torment would last, when suddenly all movement stopped. I looked around and nearly jumped with terror. The simple corridor became a cave!

A natural cave, not a man made structure.

Stone, dirt, and natural formations were all around me. It smelled damp and filthy, water were oozing from some of the walls, and the light was dim. I did not like the place.

The Master told everyone except me to stand in a crescent shaped row, and hold hands. He ordered me to stand behind the crescent, and not to touch anyone. I was hurt. I felt neglected, as if I were not part of the group, until one of the people turned to me and said kindly, "Don't be upset, my boy. This is for your protection." So I just stood there behind the people, feeling silly in my long white robe, but not unhappy anymore.

At that moment, Taj made a sweeping motion with his hands and body, and screamed a few words. The horrible sound he emitted was not human. It was very likely the loudest sound I had ever heard.

He continued to move his hands violently, grabbed some dirt from the ground, and threw it up in the air. He pronounced a word that to me sounded like a name, and followed it by the word "*Eehdar!*" three times.

Then he said, "*Oodkhool*," three times. Immediately, a rubbery kind of form moved to the left, changed to a paste-like substance, and attached itself like glue to the wall. The sticky, pale mess looked like ectoplasm.

Taj repeated his actions a few times, manifesting a new ectoplasmic manifestation on the wall with each call. Then, he looked at the Cheik and said, "*Tamam!*"

The Cheik and Taj were engaged in a conversation in low voices. They seemed to be in agreement, since the Cheik said, "Yes, go ahead." Taj advanced toward the ectoplasmic forms, put his hand in his jacket's pocket and took something out, and gave some to each of them.

At this moment, the Cheik stepped forward, ready to take over, and said "*Iibriiz!*" The forms burst into flame, which burned the ectoplasm and produced a thick fog. From the fog appeared human forms, but there were only six of them. The Cheik said "*Wawsabeh!*" The Master came forward, stood by the Cheik, and the Cheik repeated the word, adding, "*Anna a'mooree khum!*" and the seventh creature came.

Later Taj told me that these Afrit were originally created by the Cheik for a reason, as they usually are, and in the normal state of events were supposed to become the Cheik's loyal servants. However, the Cheik made a mistake and did not perform the exact requirements needed in the procedure of the creation, and therefore he lost control over the Afrit.

The result was disturbing. The seven Afrit developed independent and rather evil habits, and did not quite obey the Cheik as they should. For some reason, the only one who could call them to appear was Taj.

However, that is all he could do. Since Taj was not an Enlightened One, he could not control them once they came, and to a certain extent was at their mercy and had to have an Ulema present if he were to avoid potential harm.

As for another Ulema controlling them instead of Taj and the Cheik, that was not possible. The Ulema have four categories, based on their form of existence. Some Ulema are physical and live as humans, like the Master and the Cheik. Some used to be physical, but were no longer so.

Some, like the Tuareg, straddled both forms. Others have never occupied a human form. All four versions of the Ulema can exercise immense powers, no matter if they are physical or non physical, but a physical Ulema can only control non physical entities, such as these Afrits, if he was their creator.

I shuddered as I watched the Afrit. At this point of my studies, I had my share of supernatural incidences, but I have never been so shaken before. In the semi darkness of this miserable, damp place, the Afrits were truly terrifying.

Each had a more or less human face, but in this almost normal face the eyes were not at all normal. Instead, each Afrit had two circular orbs, with white background and a black pupil that stood out as if painted. The eyes did not move.

If the Afrit wanted to look to the side, it had to move its whole head. The head was not connected to the body. Instead, it floated in a disconcerting, eerie fashion, just above the body. When the Afrit manifest, their bodies often appear first and for a few minutes appears headless, until they choose to manifest the head.

This fact, coupled by their appalling ugliness, can frighten a human being to the point of death. There had been recorded incidents of people dying of heart attack or stroke caused by such events.

I kept myself as calm as possible and continued to study the Afrit. The heads were bad enough, but the bodies were even worse.

They were tinted a shadowy, ugly, dark color. The torso resembled the shape of a bat. Their arms were attached to the back of the body, and the hands had extremely long fingers. Since the Afrit don't eat or breathe, they don't need a stomach and a diaphragm.

Therefore, the body had a sort of visible cavity in the front, where these organs would have been. The legs were twisted, like entangled wires, which must help the Afrit as they jump. They rarely stay in one place for long, and keep shaking and moving and twitching.

They looked back at us, their ugly faces twisted in a devilish, vicious smile. They kept chattering among themselves and pointing at us with their long fingers.

But Taj told me that despite their apparent boldness, they were afraid of the Enlightened Ones. Any Afrit can see the shining auras of the Ulema, and for some reason they are terrified of these auras.

The Cheik commanded the Afrit to open the door. I did not understand the language he spoke, but I figured it out because he used the word "*Babu*," which is so similar to the word *Ba'ab*. Babu is really a door, though, while ba'ab is a gate, but the words were close enough to make it clear to me that they were going to open the door to the underground world.

I was speechless with anticipation. Everyone stood still, looking at the far wall of the cave, so I stared at it too, not knowing what to expect.

The far wall of the cave suddenly collapsed, in total silence. It felt like a silent movie, because there was no dust and no sound of falling stones during the procedure.

The stones tumbled down quietly, one by one, disappearing altogether rather than forming a solid pile.

The wall was replaced by dark, hazy fog, that allowed us a glimpse of some far away buildings. "Now," said the Cheik to Taj, "Let's follow the Afrit, but don't let them play tricks on you."

Taj nodded. We went through the fog, following another corridor and crossing identical rooms that seemed to follow each other in succession, all the while seeing the far off buildings in the distance.

The Cheik started reciting something.

The Afrit were jumping up and down like carousel horses, while pushing forward with great speed, and were already a good distance away from us, going on their own mysterious errands.

Taj said to me, "You can now move to the front, it's safe now, the Afrit won't pay much attention to us anymore." I quickly moved near the Master at the head of the line, and no one took notice of what I was doing. We did not move on yet.

The Cheik asked Taj to show him a piece of paper he was holding, probably a kind of a map, and asked, "Do you know which room we need?"

"Yes," said Taj. "I know exactly where it is, it's very near us. I will go in, and if I find something, I will bring some pieces back to you so you can see them, and then we can all go in and bring everything."

Taj left for about five minutes, and returned with a beautiful pearl necklace, a few diamonds, and some Phoenician coins. He told the Cheik and the Master, "We can go in now, but remember, you promised that all the gold belongs to Taj."

"Of course," said the Cheik casually. "But remember," said the Master, "We are not just going into the treasure room. You will also take us to the other room, as you promised."

It was clear to me that the Ulema were not in the least interested in the treasure, but there was something else in this underground cavern that meant much more to them than any gold or diamonds.

The Ulema do not need gold. They can manufacture whatever wealth they need, and they never manufacture or acquire more than they need. Riches are of no interest to them at all.

"Certainly I will take you to the other room," said Taj. "I know exactly where it is." He seemed quite pleased by the bargain.

We followed Taj into a small, closed room. It had no windows but was brightly lit, allowing us to see gold, gems, diamonds, and pearls stashed in boxes, jars, or simply thrown on the floor in heaps. However, I was not very interested in gold either.

What I wondered about was the source of the mysterious illumination. No windows, no lamps, no candles, but bright light in every corner of the room. What could cause this?

Suddenly I realized it had to be the same type of light that was discovered in the Pharaonic tombs and catacombs of ancient Egypt.

Originally, the archaeologists who went there were baffled by the light in the Egyptian tunnels, until they discovered the contraption that the ancient Egyptians had created. They found conical objects that functioned like modern batteries, producing light that was so much like normal electrical light that there was hardly a difference. The batteries had to be placed in a certain way against each other, or they would not light, and worse, could burn the user since they packed a lot of energy in their structure. I suspected this had to be the same type of illumination.

Taj pointed the door that would take us to the room the Master wished to visit. The Master asked him, "Do you want to come with us?"

"I will follow you as soon as I am finished here," said Taj, grinning. He pulled some linen bags from under his jacket, and busily started filling them with the treasure.

The Master smiled indulgently at him, as if Taj was a child playing with some toys that meant little to adults but pleased the child a great deal. He said to the rest of us, "Well then, let's go to the next room."

We opened the door. Inside it was pitch black, but the Master stepped in without the slightest hesitation, and we followed. I envied his confidence. As far as I was concerned, how did we know an angry Afrit was not waiting for us?

But since no one else showed any fear, I went with them. We could see nothing, but the Master kept talking to us and so we were able to follow him. All of sudden, bright light filled the room.

I blinked a few times, and then saw the Master standing by one of the walls, holding two conical, golden objects in each hand, positioned against each other. I was right, here were the ancient batteries.

The room was empty of furniture other than a beautiful wooden table, carved into arabesques, much like Moroccan furniture.

The Master placed the batteries carefully on the table, making sure the alignment allowed them to continue to produce light. I looked around. Other than the batteries and the table, the only object in the room was a large Phoenician urn, standing in one of the corners.

"We are going to leave you here for a short while," said the Master to the group. "The Cheik and I are going to get the materials we need for our project."

"We'll be right back," added the Cheik with what seemed to me rather misguided optimism. There were Djinn and Afrit here! Wasn't anyone concerned about these devils?

The Master and the Cheik walked to the end of the room, very slowly, with measured, matching steps, as if choreographed. Then they reached the far wall, and literally went through the wall to the other side.

I was not exactly shocked, since I have seen the Master go through walls before. It is an interesting phenomenon, but not as mysterious as one might think.

To put it simply, the Ulema know how to control molecules; the Master had explained it to me thoroughly. Everyone knows that there is plenty of empty space between the molecules of any matter, and the Ulema make use of that fact with a specialized procedure.

As the person who wishes to cross approaches the wall, the wall gradually becomes soft, as if its molecules fragment themselves, and the human body simultaneously does the same. The spaces between the molecules of both grow and readjust. The person and the wall keep their shapes for an instant, then their molecules mingle and allow the passage.

At that moment, the person passes to the other side, the molecules separate, and both wall and person become solid and normal again.

The rest of us waited for about half an hour. I was beginning to worry. The Cheik said they would be right back! Something must have prevented them from doing so.

Perhaps the Afrit, who has by now completely disappeared, took them away, kidnapped them, led them somewhere horrible? I asked some of the other people if they knew what was going on, but they had no idea where the Cheik and the Master went.

However, they did not seem worried, making it clear to me that they trusted these two to know what to do. "Don't worry," one of them said to me. "They can handle a lot worse than those stupid Afrits."

"I don't wish to contradict, Sir," I said, "but these Afrit seem pretty dangerous to me. The way they were pointing and smiling..." The others laughed. "I have seen the Cheik and the Master handle much worse entities," said the man who spoke to me, very kindly. "Remember, the Afrist are cowards. They are mortally afraid of the auras of the Ulema."

"But I understand the Cheik needs some help because of the way he handled their creation," I said.

"Yes, this is true," said the man. "These Afrit did turn out a bit wild. But with the Master there, they will never dare to harm them." I had to be content with that. So I went in search of Taj, to see how he was doing with the treasure, perhaps help him finish filling his bags.

I called him and was about to reenter the room, but I heard him scream, "Don't come here!" and he tumbled out of the room, bleeding, and slammed the door behind him. "The Afrit beat me," he gasped. "Beat me very badly."

"But Taj, you could handle those seven Afrit so well! What happened to give them power over you?"

"Seven? Are you joking? There is a colony here, something like forty of fifty Afrit, and they all rushed at me and would not let me take the gold."

"Is it their gold?" I asked. "What do they want it for, anyway? They don't need money."

"No, it's not their gold. It used to belong to the Phoenicians, and now it belongs to no one in particular. But the Afrit like to play with it. They like shining things."

"But you are holding one bag, I see."

"Yes, I managed to save one bag. They got all the others, those slimy devils." He smiled, regaining his composure. "Never mind, though. After all, I will be a very wealthy man even with just one bag. This treasure is amazing... Anyway, we must secure the door. Hold the bag for a minute." He pushed the bag in my hands, turned, and repeated the same words he used when he originally called the Afrit, and gestured in the same way.

While he was doing that, I heard shrieks and screams, which he later explained was the way the Afrit spoke as they were chased away. "That is that," he said, surveying the door with satisfaction. "They won't bother us again." He took the bag and smiled at me through the caked blood and filth on his face. "A successful treasure hunt, ah, Germain? And some day I'll come back for more."

Back in the other room, I saw, to my considerable relief, that the Cheik and the Master have returned. The Cheik was holding a stack of forty or fifty sheets made of shiny plastic, or plasma, or glass, and the Master had the same size stack, but of a different type of material, brownish yellow like corn.

"What is that?" I asked Taj.

"I have no idea," said Taj. "They only told me which room I was supposed to take them to, but they did not tell me what project they were engaged in. I must say

I have a hunch it is something terribly important." I thought so too, since the Cheik and the Master seemed to be extremely solemn, and everyone else was completely silent. There was a strong feeling of expectation in the room. They each put his stack on the table, the Cheik on the right, the Master on the left, leaving a space between the stacks, and I noticed that the space matched the size of the stacks.

The Master brought the urn from the corner to the table, and made a motion of pouring something out of the urn into the space between the stacks. I saw nothing coming out of the jar, but I figured that it might be an invisible substance.

This went on for about twenty seconds, then the Master returned the urn to the corner. The Cheik took one sheet from his stack, and put it in the space between the stacks. The Master then took a sheet from his own stack, put it on the Cheik's sheet, and waited a couple of seconds.

Then the Master flipped his sheet back side up, and to my absolute amazement, there was print on the sheet, strong and black, consisting of strange symbols and letters I did not recognize.

Piling the sheets on top of each other, they did the same to all of them. Surprisingly, the stack, when finished, was reduced in size to about a half of the original sheets, even though I could not see it reducing itself while it was worked on. I think that the plasma sheets were absorbed into the corn-like paper as the print was produced, but I am not sure. The Cheik pulled out a silk scarf from his robe, put the stack on the scarf, rolled it, lifted the ends of the scarf and tied them together, all in a ritualistic way. Then he said, "*Al Hamdu*" twice.

They turned to go, and we left the room. The Master, throughout the entire time, paid hardly any attention to me, which bothered me a little. I felt neglected, even abandoned. He must have noticed my unhappy face, because he put his hand on my shoulder, took me back into the room, and said, "Look!" To my amazement, the room was entirely empty. The table and the urn had disappeared.

I was confused and uncomfortable.

I could not understand why all that was necessary.

Why Afrit? Why those doors?

Where did the table go?

What was this document and why was it worth all this effort? He laughed at my questions and said, "Look at the wall." The light was dimming as we spoke, and finally disappeared. It seemed this adventure was over, and I said, rebelliously, that I wish things were made clear to me, because otherwise, I have learned nothing.

I will explain everything later, Germain. I promise"

"But what about the city you said we are about to see?

The city where the Founding Fathers of the Ulema used to come to?

The city from before the Deluge?"

"So you want to see more? This was not enough?"

"Yes," I said. "Basically, all I saw was you and the Cheik going through a wall and Taj fighting with the Afrit, which I admit were scary but were not too significant, I believe. I did not see anything remotely connected to the ancient city."

"Well," he said, "in this case, turn, and walk with me. You are already walking in this city."

I looked around, and saw nothing, but he said, "Keep walking, it will come."

I should have trusted him more fully. After all, when did he ever disappoint me? I felt remorseful as the miracle began to enfold in front of my eyes, but thankfully, he did not hold my short term rebellion against me, and went on cheerfully enough.

Slowly, the ancient city started to appear like a Polaroid picture in front of me. The colors of the city were such as I have never seen before, glowing colors of incredible beauty.

The Master explained that this was because the city was located in a space that had the same temperature everywhere, and no pressure on any object. Unlike earth.

"What do you mean, Master, when you say 'unlike earth' like that? Are we not on earth?"

"No, we have left earth when the Afrit opened the door and made the cave wall collapse. We are now in another dimension," said the Master. "Everything looks a little different here." The city became clearer, and I thought it looked like a holographic projection, either from the past, or from the future.

The buildings, though beautiful, had a sense of alien, remote places.

We were now walking in a well-illuminated street, the windows of the buildings shining with lights as well. The air was soft and fragrant.

"I see buildings and streets," I said. "But where are the people?"

"They are here, but they are invisible to you. Your eyes are not constructed to see them, not yet," he said. "Well, it is time to leave. Let's go up these stairs." We started climbing a very high, stone stairway that led from the street into a destination that was not quite visible.

I was surprised that we were not retracing out steps into the Cheik's house, but the Master said there was no need for that, and that exits were available in various locations, and not as difficult to achieve as entrances. So we climbed the stairs, and when we reached the top, I saw a huge gray wall on my left, and noticed that the pavement turned into sand.

The huge gray wall was the side of the Anunnaki stone.

I understood that we exited from a hole under the big stone, were out of the strange dimension, and back on earth.

"So that is what Taj meant when he said he would make the stone fly?" I said.

215

"Yes, a rather poetic way of describing our trip," said the Master.

"Master, I am not wearing the white robe! I am wearing the normal clothes I left at the Cheik's house."

"Indeed, and so is everyone else," he said, pointing to the rest of the company, who were already standing near the giant stone, and wearing normal clothes.

"So what did we come here for? Surely not just to give Taj his treasure?"

"We came for the book, Germain. Everything we did was much worth it, even the encounter with the unpleasant and stupid Afrit. We have recently heard that the book was here, in this dimension, after having searched for it unsuccessfully for generations. And now we have recovered a copy of the most important book in the world."

"The strange book you printed from the stacks? What is it?"

"It is one of the very few copies in existence of what is probably the oldest book to have ever been written. A book the Anunnaki had valued very much. It is called *The Book of Rama Dosh.*"

I didn't know why, but a shiver went through my spine when I heard the name of the Ancient book; the sound of the name triggered a reaction in my mind. For a second I had a feeling of tottering on the brink of a dark, warm abyss that contained something older than the universe, and glowed with endless stars. It passed quickly, and the Master continued.

"In the future, you will have the privilege of studying it. It contains the knowledge that may, some day, save humanity from its own folly. At least I hope so with all my heart. And now, back to Damascus! Our friendly driver is waiting for us in the car."

***   ***   ***

216

117. Humans' Link To The Beginning Of Everything, The Anunnaki, and "God"
⌘ ⌘ ⌘

Introduction
On the origin of Man and the existence of God
From the lectures of the Ulema de Lafayette and Ghandar
Kira'at, part one
Kira'at, part two
Kira'at, part three
Ulema Ghandar said
Kira'at, part four:

Introduction:
On the origin of Man and the existence of God

- The origin of "Man" (Human Races);
- The first bio-organic creation of quasi-humans (half animals-half humans);
- The early 37 or 47 different human species;
- How, when, and why their bodies and minds developed;
- Where did they live;
- The existence and role of "God", or "Gods Creators" in the genesis, life, and development of humans;
- The extraterrestrials races who created and/or shaped the early humans;
- The reasons for upgrading already existing humans on Earth;
- The Anunnaki-Igigi-Modern humans equation; and finally

217

Worth mentioning here, that Ulema de Lafayette used a very simple language, far from jargon and complicated words and expressions, to explain to their students, adepts and novices, all these confusing and equally challenging subjects.

Yet, in their simplicity and easy-going narratives, the Ulema used some key-words that only contemporary science, and particularly modern forensic anthropology, and quantum physics could explain to an advanced modern mind, the enigmatic mysteries, and incomprehensible creation/evolution/genetic development of the human race.

From the lectures of the Ulema de Lafayette and Ghandar:

a- **Kira'at** (Reading/lecture), number 3,654.
b- **Jalsa** (Meeting; gathering), first, part one.
c- **Fasel Ketab** (Chapter or Section of the Book), number 87.
d- **Dirasa** (Subject; study): Near future.
e- **Donia** (Life; universe): Part one.
f- **Ulema-Oustaz** (Teacher): Honorable Master Ghandar.
g- **Markaz** (Place; location): Benares, India.
h- **Zaman** (Time; year): 1963.

⌘⌘⌘

Kira'at, part one:

- Don't feel threatened or humiliated by the expression "as it has been decided upon by the Anunnaki."
- There is no reason whatsoever, to think that we are the slaves of an extraterrestrial race.
- Sometime, perhaps always, some major decisions by a more intelligent, advanced and responsible group or an elite are necessary, to keep order, peace, and justice in the universe, in our lives here on Earth, in our societies, and even in our relationships with others.
- Otherwise, our lives, the fabric of our societies, and the stability of our very existence will be seriously

218

jeopardized, and chaos will disrupt all forms of progress, happiness, and success in our endeavors.

- Consequently, decisions must come from those who are superior to us, from those who are more intelligent than us, from those who really care about us, unconditionally, no strings attached.
- So far, humans were/are not able to do so.
- Of course, some great minds have accomplished a lot. But if you revisit history, you will find out that some of the greatest minds were either ridiculed or murdered, or even forced to take poison!
- As long as, here on planet Earth, we have markets' competitions, different understanding of what is right and what is wrong, as long as we have so many different religions, each one claiming to be the most truthful religion, and as long as, using aggression (in any form or for any reason) to convince others, to intimidate others, to rule others, and to protect interests at a personal level or at a state level, as long as we are controlled by these rules and differences among us, the human race will remain negatively affected, greedy, violent, selfish, materialistic, and unable to seek and reach happiness, peace and the ultimate truth.
- Only the Anunnaki are capable and entitled to maintain law, order, peace and stability for us, for our families, for our children, and for our future.
- The Anunnaki know best, because they have created all life-forms on Earth; from the smallest fish in the oceans to the biggest idea or concept a human being can think of.
- Materialistic things, including ideas, concepts, dreams, joys, fears, and aspirations, all of them were originally created and propagated in our minds, bodies, and actions, at the time the Anunnaki conceived a "picture", a structure, a substance, and a permanent physical-mental nature of the human race.
- They created us thousands of years ago, long before Man used the word "God", and followed any religion.
- Here, I am talking exclusively about the Anunnaki, and not about other extraterrestrial races, and they are so many all over the universe.

- When it comes to our place, role and future on the landscape of the universe, and especially our relation to the Creation, Creator or Creators, life after death, and our possible existences in other worlds, there is a big and primordial difference between the Anunnaki and other extraterrestrial races.
- Because the Anunnaki have shaped us physically and mentally, they looked upon us as their cherished and favorite creation.
- Here, there is a direct relationship between us and them.
- In fact, a part of us resembles a part of them. And we shall talk about this resemblance very soon.
- On the other hand, extraterrestrial races who had nothing to do with us at any level, and for any reason, saw us as an inferior species; inferior in so many ways. Thus, these "stranger extraterrestrials" care less about us, about our families, and the continuity of the human race.
- In the ancient past, these "stranger extraterrestrials" interacted very briefly with early forms, early quasi-humans, and early "categories" of human beings.
- Nothing great came out of it.
- They did not even bother to teach us anything.
- Would you teach an insect how to write a symphony, how to build a hospital, or how to plant a rose garden? No, you would not, because you believe that an insect is incapable of learning and creating such beautiful and intelligent deeds.
- Well, this is how exactly, many extraterrestrial races viewed our ancestors, and ironically continue to think about all of us.
- We are not worthy of anything, because on the galactic map, humans have so little to offer to the cosmos. This is not a negative attitude.
- This is not pessimism. This is a fact.
- But, the Anunnaki do not share the opinions of other extraterrestrial races.
- To them, we, the human beings mean a lot, but to a certain degree.

- But here, I must be totally honest with you, dearest friends and students.
- Look around you, everything you see has a past, a present, and a future.
- Look at all these beautiful plants in the fields. Their past is seeds. Their present is what you see. Their future is to fertilize the earth, or to become something else by changing into something new, different, another organic substance, so on.
- Well, this was not the case for humans, when humans first appear on Earth.
- There are so many opinions and ideas about our origin on Earth. And we will not succumb to rhetoric.
- Some of the early human races, and quasi-human races vanished for many reasons.
- Others survived and relatively developed into a more intelligence species.
- Others became us, no matter what science calls us.
- The human category that has survived and became us, is essentially a sort of a genetic animal-human blend.
- Yes, we were, and still are a genetic formula.
- And as you know, a formula does not need a past to exist, only an idea at the beginning, followed by research and experiments, leading to the creation of something new.
- This is exactly what happened to us; how and why the early forms of humans came to exist, and develop into something else, without a definite and a well-prescribed future. And I will explain this, because it has a major impact on our future.

Note: A pause for 5 minutes.

Kira'at, part two:

- Before I continue with the Dirasa, I want to explain to you, briefly, something very important. And keep it in your mind.
- Everything you see in the universe came from something else.

- Not necessarily from the same material or the same substance.
- Nothing in the Donia (World) appears for no reason.
- Nothing in the Donia, comes from nothing, except the beginning of the universe itself.
- And I will talk about this, later.
- Always, and always, keep in mind, that there is an origin for everything.
- And this origin, or as we call it "Source" is also what we call in our study "Khalek" (Creator).
- Creator is in fact "Creators".
- In the weak mind of humans, what appeared to be (To our ancestors, prophets, and creators of our religions on Earth) the real, unique, and only God, was in fact, the "strongest" and most powerful god from all the other gods (Humans came to know, or heard about) who managed to outlaw other gods, to eliminate other gods, and convince our ancestors, that He is above all gods; the only one; and the creator of human beings, and the universe.
- Struck with fear, and limited in knowledge, science, and history of the universe, our ancestors (From all the nations, and different religious beliefs) accepted that "god" as the real god of the universe, and creator of the human race.
- The Habiru (Israelites; Hebrews) called him Yahweh.
- The Phoenicians called him El; Baal.
- The early Christian Arabs and Muslims called him Allah.
- The English speaking nations called him God.
- The French called him Dieu, so on...
- The truth is, the "Origin", the "Source", and the "Creator" of humans, and life as we know it today, are in fact one and the same: Creators.
- The word "Creators" means, those who generated life on Earth.
- In other dimensions, they are called differently, but we are not going to worry about that, for now.
- The "Creators" were a very advanced extraterrestrial elite who used to travel many parts of the universe. A more

222

accurate word would be the "Multiverse". And Earth is one "layer" or one single dimension of the "Multiverse".

- Thus, it is extremely <u>important to believe in a</u> "Creator", whether in the singular or the plural form.
- You look at your wrist-watch, and you instantly realize, that there is a watch-maker.
- You look at the car you drive, and you understand that there is an automobile factory.
- Thus, when you look at your body, you should understand, that there is a body-maker.
- Very good. Then, let's start with this body of yours, you are now looking at.
- This body of yours is made from cosmic dust.
- Everything you see on Earth, and around Earth is made from a cosmic dust.
- You can call this cosmic dust whatever you want; Cosmic rays, gases, molecules, particles, carbon, proton, neutron, atoms, bubble, so on.
- The cosmic dust was produced from the collision of several layers or space bubbles at the beginning of time and space.
- Part of this cosmic dust fell on Earth.
- The cosmic dust on Earth produced all sorts of things; mountains, rocks, strange and archaic life-forms, plants, clouds, and yes, the primordial and early forms of creatures that moved or walked on five legs, four legs, three legs, and without legs.
- Some were animal-plants; some were animals; some were something else; and others animals and animals-humans.
- All of them were produced and came from the cosmic dust of the exploded universe.
- At that moment in history, time and space, and beyond, the "Creator", the "Creators", and the God we know had nothing to do with humans, early humans, and animals-humans life forms.
- In other words, the human race (From its early archaic form to the present) was created by non "Creators", "Creator",or a "Judeo-Christian-Muslim God".

223

- And this applies also to all the extraterrestrial races and their habitats.
- The extraterrestrials like us, were also created by and from a cosmic dust.
- But they are different from us, physically, emotionally, psychologically, and mentally. And we will talk about this, some other time.
- So, when Earth was created, and during the very early existence, and following stages of Earth, some forms of animals and humans were created too.
- The early humans had a multitude of categories.
- Each category was shaped differently.
- Every shape depended on its origin/place of creation.
- For instance, humans or quasi-humans who came from the sea, looked very different from their counterparts who came from land.
- Some humans or quasi-humans were reptilians. Or at least quasi-humans, with the face of an archaic undeveloped human, with a long tail.
- Some walked on three legs.
- Some walked on four legs, and so on.
- No intelligence faculty was yet installed in those bodies.
- Call them robots or moving creatures if you want.
- They were a blending and a matching part of the ecology and landscape of Earth.
- And since the "Conduit", or an "intelligence faculty" was not an original part of their organic or bio-organic substance (Structural composition), a future or a reason for reproduction was not necessary, even though, in many instances, materials, substances, and species can reproduce without intelligence.
- But this is what makes the primordial and definite difference/distinction between humans and other living-forms (Life-forms).
- At that time in history, 37 different quasi-humans (half animals-half humans) lived on Earth. Some esteemed Ulema suggest 47 different species. We will talk about that in our next Dirasa.
- On other planets, in different galaxies, in distant dimensions, life evolved differently.

- Different kinds of life-forms were created.
- Some life-forms were the extraterrestrials.
- And some of these extraterrestrials were or became highly advanced. Not all of them.
- Some extraterrestrials who lived on planets similar to Earth, looked like modern humans, but not totally. Some were giants, others were short, and other groups were horrible-looking creatures.
- The only three races who shared some similarities with us were the Lyrans, the Anunnaki, and the Igigi, who were mentioned in the Book of Ramadosh, in the Bible, in the Book of Enoch, in the Phoenician Cosmogony, and the Babylonian clay tablets.

Note: A pause for 5 minutes.
Ulema Ghandar said:

Kira'at, part three:

- Between 450,000 and 460,000 BCE, huge Anunnaki's spaceships hovered over the Earth.
- First, they flew over Madagascar, Brazil, Australia, and Central Africa.
- Then, they began to scan the lands in the Near East.
- An immense mother-ship will stay still in the air, some thousands feet far from the surface of the Earth.
- And a great number of small flying machines would exit from under the belly of the mother-ship, and head toward Earth.
- They were piloted by Anunnaki.
- Their missions were to probe Earth, scan the lands and waters, and locate certain minerals, natural resources, and some rich aquatic substances found only in the seas and the oceans of the Earth.
- The Anunnaki accompanied by their allies, the Igigi landed in Phinikya (Phoenicia; Modern Lebanon).
- The huge mother-ship remained in the sky and began to orbit Earth.
- The small spaceships landed on Earth.

- The mother-ship also functioned as an intermediary space-station between Earth, the Moon, and Mars.
- Bear in mind, that before visiting Earth, the Anunnaki have already established colonies on the Moon, and on Mars.
- Some traces of those colonies, and minor evidence(s) of the pre-historic Anunnaki's civilization, are still visible on the Moon and on Mars.
- As soon as the Anunnaki and the Igigi landed in Phinikya, they began to colonize the area, and build settlements.
- The settlements consisted of movable (Mobile) prefab living units and quarters. You can call them if you want prefabricated homes.
- The chief of the Anunnaki who was in charge of the colonies in Phinikya was called Aa-kim-Lu. His other names were Anoon Elah-Im; Anu. Il-Ohiim; Anu-Ela-Kim.
- Some of the earliest great findings of the Anunnaki and the Igigi were the Maha'rit; The Ourjouwan, and the Zaa'-faran.
- The Anunnaki and the Igigi spent many years in Phinikya; they built huge cities, housing facilities, and laboratories.
- Then, they found out that Earth was extremely rich with natural resources, and quickly realized that they needed a larger team of workers to mine and extract these riches.
- Years ago, while exploring other regions on Earth, the early Anunnaki expedition discovered heavily built creatures in Madagascar, Brazil, Australia, and Central Africa.
- Some of these creatures were beasts, some half humans-half animals, and others were reptilians.
- Here, I have to remind you that the Anunnaki did not create these creatures. They were here on Earth, long time before the Anunnaki and the Igigi descended on Earth.
- So, the Anunnaki went back to Madagascar, Brazil, Australia, and Central Africa, looking for these creatures as potential workers.

- You already know what happened next.
- They captured those horrible looking creatures, and tried to domesticate them, exactly as humans did, when they captured wild horses and dogs.
- So, they caught them in masses, and brought them to Saydoon, Tyrahk, Kadmosh, Adonakh, Ilayshlim, and Markadash, their colonies in the Near East.
- Those archaic creatures were not intelligent at all.
- Consequently, they were unable of fulfilling any task that required intelligence, or even a minimal understanding of what they were supposed to do.
- The Igigi decided to create (Manufacture) creatures capable of carrying and executing hard labor and continuous physical work.
- Well, what they created was not what they have hoped for. In fact, their creatures looked more horrible than their predecessors, and acted and moved like deformed beasts.
- In short, the Igigi's creation experiments were a total failure.
- Then, the Anunnaki interfered, and after many attempts, trials and errors, and new genetic experiments, they succeeded in creating a satisfactory breed/species of humans.
- To accomplish this, they had to slaughter an Igigi, take his blood, mix it with earthly elements, blend the whole thing with wet Tourab (Clay), and breathe into the mold.
- This genetic procedure was one of several, and varied genetic experiments and methods. Because, the Book of Ramadosh and the Akkadian clay tablets tell us about other means processes, and elements, the Anunnaki used to create humans.
- Some experiments were conducted in Chimiti.
- In Akkadian/Sumerian, Chimiti means tube; containers. In other words, laboratory's tubes and containers.
- Some genetic experiments required the impregnation of Anunnaki women (Goddesses).
- Some experiments required the semen of Anunnaki men (Gods).

- Some experiments required upgrading an existing quasi-human race. So on.
- The Book of Ramadosh, as well as the Akkadian clay tablets gave us the names of the Anunnaki men and women who created our ancestors.
- But up to that moment in history, the early creatures manufactured by the Anunnaki remained at a robotic level.
- Years later, the Anunnaki upgraded their creatures by adding 13 faculties in their brain.
- This addition allowed the early humans to communicate with their creators.
- These faculties are called Fik'r-ra-ma. Meaning a bulk of understanding and reacting.
- Years laters, the Anunnaki added some (Unknown to us) aquatic bacteria such as fungi, much needed for the development of certain organs of the early humans.
- The Anunnaki kept on exploiting Earth, building more colonies, and adding new cities neighboring Saydoon, Tyrahk, Kadmosh, Adonakh, Ilayshlim, and Markadash.
- Among the new major cities were Baalbeck, Anfa, Sippar, Nippur, to name a few.

Note: End of the Kira'ats for that day.
Part four of the Kira'at resumed on the following day.

Kira'at, part four:

- The Anunnaki and the Igigi stayed on Earth for hundreds of thousands years.
- And then, for reasons we do not totally understand, they abandoned their operations and cities on Earth. At least, in Phoenicia, Babylon/Mesopotamia, Anatolia, Turkey, and surrounding areas.
- Then, years later, the Anunnaki and Igigi retuned to Earth.
- But this time, the Anunnaki decided to create a more advanced human race.
- And they did.

- They called that race Bashar.
- This happened around 300,000 BCE.
- Around 125,000-100,000 BCE, the Anunnaki created the early human women, and were called "Women of the Light"; they were the early female-forms on Earth.
- Thousands of years later, the people who lived in the Arab Peninsula and the lands bordering Persia, the United Arab Emirates, and India, began to call these women "The Women of Light", and those who were allowed to "mix with them" were called "The Sons or Children of Light".
- From this early human race, all humans came to life.
- The Judeo-Christian God had nothing to do with the creation of the human race. In other words, the God we know, revere, and fear today did not create us. Even the word or term "God" did not exist in the early stages of the existence of the human race on Earth.
- Around 65,000 BCE, the Anunnaki created the final form (Physically and mentally) of humans.
- This final form did not change much throughout the ages. Basically, it remains the same to the present day.
- For reasons we know, and reasons we don't know for sure, the Anunnaki got mad at the human race, and the supreme god/lord of the Anunnaki decided to wipe out the human race, from the face of the Earth.
- The Anunnaki mythology tells us that the Anunnaki created avalanches of Tsunami (Great Flood) that changed and altered the geography and locations of lands and mountains on Earth, and killed almost everybody.
- Some humans were saved, because a group of Anunnaki elite interfered on behalf of our ancestors, and warned them of the forthcoming annihilation plan of the supreme god/lord of the Anunnaki.
- Earth was repopulated, and new cities were erected in many regions of the Near East and the Middle East.
- This, included the lands of the Persian Gulf, Hadarmoot, Bahrain, Qatar, the Arab Peninsula, Turkey-Anatolia, Mesopotamia-Babylon, Phoenicia, the Fertile Crescent of Syria, Jordan and Palestine, etc...

- From 65,000 BCE to 7,000 BCE, many things happened on Earth.
- But what really interests us here, is the emergence of a new human race.
- This emergence occurred around 5,000 to 6,000 BCE.
- At that time in history, and for the first time in the annals of humanity, a human race with a developed mind, knowledge of arts and science, a capability of recording historical events, with tools to erect cities and develop agriculture, apparently surfaced from nowhere, out of the blue, and without no direct link to the early humans.
- No anthropologist or a theologian seems capable of explaining the missing link between this new human race, and the anterior ones, and the sudden development of civilizations on Earth.
- Pyramids were built. Extraordinary cities like Baalbeck were erected in no time. Immense plantations flourished everywhere. Coherent languages were heard. Laws were formulated, to govern societies, city-states, and far distant nations. Organized religions, pantheons, deities, and worship rituals appeared everywhere. And no historian or an anthropologist succeeded in explaining the sudden appearance, and origin of humanity's culture, languages, and civilization.
- It was at that time in history, that finally the Anunnaki got their act together, and decided to create a new human race, capable of functioning, reasoning, and understanding what is right and what is wrong.
- It was at that very specific time interval, that the first day of humanity's history was recorded. Because, that was the moment, when the Anunnaki created us in their image (Final creation) using their own DNA.
- This explains when, why, and how, the quasi-humans became overnight intelligent humans, and all of a sudden, cities, temples, art, science, music, poetry and human drama came to life.
- It was neither evolution, nor that God who asked Abraham to sacrifice his son, who created Man. But, the Anunnaki's genetic creation of Man.

- If it was God, then why it took God so long to create us? To create a reasonably looking human race?
- Science, archeology and anthropology have proved to us, that mankind was not created five thousand years ago, as claimed by the scribes of the Bible.
- The skeletons and remains of humans who lived millions of years ago were not in the image of the God of Abraham, Jesus and Mohammad.
- The early humans were half-animals.
- I don't think this "half-animals image" was created in the image of God, as claimed by the Prophets and the scribes of the Bible.
- And if it was evolution, then how can we explain the sudden "appearance" of languages and civilizations without a direct link from 7,000 BCE to the early days of humans on Earth?
- The Bible and Holy Scriptures are silent on this issue.
- Why humans who lived before 7,000 BCE could not talk normally? Could not build Pyramids? Could not write? Could not understand human anatomy? And all of a sudden, without a gradual and systematic development, and a phase-by-phase progress, they start to talk all kinds of languages, build huge cities, sail the oceans, and perfectly understand anatomy and astronomy?
- Why? Because of the immediate and sudden intervention of the Anunnaki prompted all this. They have jumped and catapulted our development in no time.
- Neither God, nor evolution had to do anything with the creation of the modern human race.
- As I told you before, at the time when the Anunnaki created the early human race, the idea or concept of "God" was totally unknown to the human beings. Thousands of years later, Abraham created the image and existence of "God". And there is a story behind his invention. Briefly, it goes like this.
- As I told you before, at the time when the Anunnaki created the early human race, the idea or concept of "God" was totally unknown to the human beings. Thousands of years later, Abraham created the image and existence of "God". And there is a story behind his invention. Briefly, it goes like this.

- Anu was the supreme and ultimate Anunnaki/Sumerian god who reigned over the heavens. He was the head of the Sumerian family's genealogy. Anu had two sons who played a major role on Earth, and in the affairs of the early human race. Those sons were:
- 1-En.Ki, also called Ea, who became the master of the Earth, the rain and waters. His mother was Antu.
- 2-En.Lil, also called Illu, who became the master of the air and winds. His mother was Ki. So, they were half-bothers. Both came to Earth and ruled over large colonies.
- Many mythologies of the Near East and the Middle East told us that half-brothers, who ascended to power, usually hated each other. So, En.Ki and En.Lil were no exception.
- En.Ki failed to mine sufficient natural resources on Earth. His failure enraged the Anunnaki. So, the Anunnaki Council decided to replace him. Their obvious choice was his half-brother En.Lil.
- The Anunnaki Majli (Council) was very pleased, for En.Lil delivered enough quantities of Earth's riches.
- Anu, the father of En.Lil and the supreme chief of the Anunnaki Council paid a visit to his son En.Lil, blessed him, and appointed him master of the Earth.
- We read in the Sumerian Genesis of the Grail Kings:
  "To Enlil the Earth was made subject.
  The seas, enclosed as with a loop.
  They had given to Enki, the Prince of Earth."
- En.Ki is disappointed. He is furious. En.Ki said: "I am the great brother of the gods. I am he who was born first son of the divine Anu."
- At that time in history, the Anunnaki were ruled by Antu, a female deity, although Anu was the central figure. The Anunnaki world is a matriarchal society. But things began to change, and all of a sudden, male Anunnaki seized power.
- The supreme Antu lost her throne, but remained a revered goddess. Anu is in charge now.
- En.Lil was not friendly toward early humans (The terrestrial species to be created soon by the Anunnaki

232

geneticists.) So, En.Lil instigated the Flood on Earth to annihilate the human race he passionately hated.

- En.Lil was also accused of destroying the entire Sumerian civilization and heritage. In short, En.Lil hated the human beings.
- Humans feared En.Lil. He was merciless. En.Lil destroyed many cities on earth, one of those cities were Ur, the homeland of Abraham.
- Abraham had to escape the fury of En.Li, after he witnessed the destruction of Ur. So, Abraham left Ur searching for safety. Any land he will find would be safer than Ur.
- But En.Lil caught up with Abraham. En.Lil did not trust Abraham, and Abraham did not trust En.Lil
- So En.Lil told Abraham that he was the lord of the land "Earth" and he expected from all humans an absolute loyalty and obedience.
- Abraham believed him, and assured him, that he will be loyal and obedient to him.
- Abraham began to call En.Lil, "Eli" and "Adoni" (Adonai), meaning in Hebrew: My Lord; My Master. In Judaic/Israelite Hebrew, "My Lord" or "My Master" became "Eli" which is the singular of Elohim. "El" means lord or god. And the "i" at the end of the word, means "My".
- And long longtime before Abraham, Jacob, Moses, Aaron, Joshua and the rest of the Biblical figures came to exist, the words "El" and "Eli" were used by the early Phoenicians, who were an offspring of the Anunnaki.
- Abraham asked En.Lil what should he call you?"
- And En.lil said to Abraham: "You shall call me Yah-Weh. For Yah-Weh is the lord of heavens and earth. And the lord is I. And I am That I Am." (Ehyeh asher ehyeh, in Hebrew: אהיה אשר אהיה)
- Then, Abraham asked his "god": "And what about the other gods who are so mighty? Should I worship them too?"
- And En.Lil furiously said: "I am your God. The only god you shall worship."
- Same scenario will be rewritten by or for Moses for his rendez-vouz with Yahweh on Mount Sinai, assuming of

233

course, that Moses did chat with God (Yahweh), and receive the Ten Commandments.

- This is an encounter hard to imagine, because it was not really necessary at all to meet with God to get the Ten Commandments, since:
- a- First, the Commandments were already written in hieroglyphic by the Egyptian priests, centuries ago; b- Second, the Commandments were already available and accessible to Moses, taking into consideration that Moses was the right hand of the Pharaoh.
- The Ten Commandments were copied from chapters in the Egyptian Book of the Dead.
- Yahweh was indeed one of the Anunnaki/Babylonian Gods. In ancient times, many scribes considered him as one of the "Sons of God", and the early Habiru (Hebrews) referred to him as "B'nai Elohim".
- The words B'nai, Bin, Ibin, and Ben mean the same thing in Hebrew, Aramaic, Syriac, Coptic, Phoenician and Arabic: Son.

Comments:
The Kira'at of Ulema Ghandar is obviously neither a detailed chronology of the Anunnaki and the Igigi on Earth, nor an integral explanation of the Anunnaki-Humans equation.
An in-depth chronology is provided in Book 2 "The Anunnaki Ulema Forbidden Knowledge: What your government and church didn't want you to know."
Worth mentioning here, that this Kira'at was given to novices, and intended to give:
1. A general idea about the obscure origin of the human race (s);
2. The role and involvement of God in the creation of mankind;
3. The incomprehensible and baffling rapid development of Man,
4. The missing link between the archaic times, and the sudden development of civilizations on Earth.

*** *** ***

234

118. The Anunnaki Christian Saint: St. Tekla
⌘⌘⌘

I. Introduction and definition
II. Tekla, the early Christianity, St. Paul, the Anunnaki, Ulema, and extraterrestrial superior beings equation
III. The early Christian legend
a-Women's leadership
b-A women's tradition
IV. The power of Tekla (Thecla) and her story
a-In the early Church
b-In the modern Church
V. Tekla in the arena with the fierce hungry animals, and her miracles
VI. The true story of Tekla, the etheric extraterrestrial being, according to the Anunnaki-Ulema

---

118. The Anunnaki Christian Saint: St. Tekla
⌘⌘⌘

In this volume, we will limit our work to two most unusual, and unknown topics to the general public; the Anunnaki Christian Saint: St. Tekla, and the crucifixion of Jesus Christ.

Please bear in mind, that the two topics discussed in this section and pertaining comments do not categorically reflect our opinions.

However, we are absolutely convinced that the accounts of Ambar Anati are truthful in their entirety.

I. Introduction and definition:

Tekla is the name of one of the most enigmatic and cherished personalities in the Anunnaki-Ulemite manuscripts, and the early literature of the "primordial" Christian Church in the Near East.

237

According to many, Tekla was an extraterrestrial being who has physically existed and lived on Earth (In Rome and Syria) in the flesh of a woman who has displayed a high level of awareness, and unearthly supernatural faculties.

This Anunnaki woman became an early Christian saint.

In Aramaic, Syriac, and Christian Arabic manuscripts, her name is written as Tekla or more precisely as Takla, and she is known to the Christians in the Eastern hemisphere (Particularly in Syria and Lebanon), as Saint Takla or Saint Tekla.

In Greek, it is written as Techla, and pronounced Cekla.

In An'akh (Anunnaki-Ulemite), Tekla's name is Hu-wera Taklaa, and in some passages of the Ulemite's literature, she is referred to as simply "Huera".

Her story was provided and religiously preserved by the Aramaic, Syriac, Assyrian, and Chaldean traditions, and Christian Syrian Church. To the early Christians in the Near East, Middle East, Asia Minor, as well as in ancient Greece and Rome, Tekla was a saint. To the Gnostics, Tekla was one of the earliest teachers and thinkers of Christianity.

To the Ulema, Tekla was an enlightened being, because she was the physical manifestation of an Anunnaki female leader.

To millions of Christians in the Western hemisphere, Tekla is totally unknown.

II. Tekla, the early Christianity, St. Paul, the Anunnaki, Ulema, and extraterrestrial superior beings equation:

There is a fascinating chapter in the extraterrestrial Anunnaki-Ulema book "Book of Ramadosh" telling a most unusual story of Tekla, who according to the early Christian churches and Aramaic traditions in the Near East, was a saint, and a female companion/disciple of St. Paul.

However, the Ulemite story is quite different from the Christian story at so many levels, because it contained passages on Tekla's extraordinary deeds, her very unique supernatural powers, allegedly bestowed upon her by the Anunnaki, and those "Circles of Lights" that appeared in the sky, right above her head, when she was thrown in the Roman arena to be eaten alive by the lions. The early Christians interpret those lights as Divine lights sent by God to rescue Tekla, while in the Ulemite's manuscripts, those lights were described as small extraterrestrial spaceships.

Also, in the Ana'kh/Ulemite version, Tekla, aka Sinhar Tekla is depicted as an Anunnaki woman, who has manifested herself as Baalshamroot, via shape-shifting, and other extraterrestrial means.

A part of the Ulema's version appeared in a book which is the sequence to "On the Road to Ultimate Knowledge", co-authored by Ilil Arbel and Maximillien de Lafayette.

III. The early Christian legend:

Outstanding author, Nancy A. Carter said: "St. Tekla is a rather enigmatic saint. There doesn't even seem to be much agreement about the spelling of her name, and at least two other places claim to be her last resting place. According to legend, she was an early convert to Christianity and a follower of St. Paul who broke off her engagement to devote herself to God. Her vengeful fiancé tried to kill her by various means, all of which were thwarted by divine intervention.

Eventually she is supposed to have hidden away in a grotto in the cliff around which the modern convent in Maaloula in Syria was built. "

a-Women's leadership:

Nancy A. Carter stated that "the Acts of Paul and Thecla is part of a Pauline tradition that provided apostolic blessing for women's leadership roles in the church. Although the events related in the Acts are legendary, a real Thecla may have lived in Asia Minor. Like many stories about Jesus and the Apostles, originally her tales were told orally. The content of the book, with its wealth of women characters, most of whom support each other (including a lioness who protects Thecla!), suggests Thecla's adventures were popular in women's circles.

An orthodox Christian, probably from Asia Minor, penned the Acts of Thecla between 160-190. The book circulated in several languages, including Greek, Coptic, Ethiopic, and Armenian. The Syrian and Armenian churches included the Acts of Thecla in their early Biblical canons. It is now a part of the Christian apocrypha. The extant manuscripts reflect masculine editing that probably de-emphasized Paul's support of women's leadership. No longer present are references to Thecla's baptizing others, which were most likely in the earliest stories. Even so, the Acts of Thecla includes a story about Thecla baptizing herself with Paul's blessing! Later Paul commissions her to return to her home town Iconium to teach and evangelize."

b- Sexual acts in the Anunnaki-Ulemite literature:

In the Book of Ramadosh, there is a chapter dealing with genetics and reproduction, more precisely, it discusses how Anunnaki give birth to their children via the "Mingling of Lights" which pass through their bodies. Ambar Anati (aka Victoria), a half-human/half-Anunnaki woman, in her autobiography said, "There is no sex...no sexual acts in the Anunnaki's community. Marriages are consumed, and children are born without a sexual intercourse, because sexual acts are considered bestial.

The human race shall remain an inferior species, as long as sex remains a necessity to humans.", Thus, the behavior of Tekla reflects the Anunnaki's concept.

IV. The power of Thecla and her story:
In the early Church:

Nancy A. Carter stated, "Without a doubt, Thecla and Paul were key symbols for the ideals of early Christian ascetic movements, especially in Egypt, Syria, and Armenia. Obviously the women's ascetic movement did not end, even though the Pastoral Epistles declared women's salvation was bearing children.

Christian ascetic practices by both men and women continue to this day. The power of Thecla's story spread throughout early Christianity.

Several early church fathers from both the East and West praised Thecla as a model of feminine chastity. She became "venerated from the shores of the Caspian almost to the shores of the Atlantic.

In the fourth century a church in Antioch of Syria was dedicated to Thecla. Another church in Eschamiadzin, Iberia, from the fifth century has a wall design showing Paul preaching to her. In Egypt, there are several paintings of Thecla.

In Rome, scholars found a sarcophagus graced by a relief portraying Paul and Thecla traveling together in a boat. At least three places claim her burial place: Meryemlik (Ayatekla) Turkey; Maalula, (Maaloula) Syria; and Rome, Italy. Tradition says that Thecla traveled with Paul to Spain. Another apocryphal Acts which mentions Thecla is the Acts of Xanthippe, Polyxena, and Rebecca (c. 270). Some women in Spain hear Paul's preaching and leave their husbands to follow him. "

V. The true story of Tekla, the etheric extraterrestrial being, according to the Anunnaki-Ulema:

The story was translated by Maximillien de Lafayette, and co-written by Ilil Arbel, Ph.D. The following is an excerpt from their Book; Chapter Seven: A Journey to Maalula, and Meeting Saint Tekla.

Here is Germain Lumiere, a young Ulema who has studied with the enlightened Ulema masters in the Middle East. He is telling the story of his relationship with Tekla, how she has saved his life, freed him from detention, and teleported him almost 2,000 years in time and space to the ancient city of Maaloula in Syria, where he met the real Tekla in the flesh. Not only as a human being, a disciple of Saint Paul, and an early Christian saint, but also as an Anunnaki spirit/entity.

Lumiere's account verbatim, as is and unedited (In his own words):

"My training usually allowed me to keep panic at bay, to look for opportunities rather than allow fear to get in the way of solutions. But this time, I was in a predicament that was, to say the least, intimidating. After beating me thoroughly, and then promising me that unless I talked, I would be executed at dawn, my captors locked me in a small cell in their infamous prison. This was not the first time I had been arrested, during the ten years since my first assignment for the Pères du Triangle, but I have always been able to get myself out, either by normal means or by the special techniques my teachers instructed me in.

This time, the normal means did not work, as was evident from my black and blue body and the cell I was locked into, and for some reason which I could not understand at the time, my special techniques seemed to be blocked by some agency that was not visible to me. No matter what I tried, I could not perform any of my escape routes, and my attempts to contact any of my friends through telepathic means seemed to be blocked as well.

At the time, at thirty-five years of age, I was not as yet at my full capacity, which one acquires only at age forty and after additional tests, but I have never failed so miserably before, and I was not sure what to do. In addition, I was in great pain, because trying so hard to escape prevented me from concentrating on healing the damage my captors did to me through their beatings.

So I sat on the floor and reflected. Since the spirit of my mother came to see me, on the day of her death, I was not afraid of dying, so the threat of execution at dawn did not cause me despair.

However, I knew I should try to go on living because my tasks on this plane of existence were not even near completion. According to my masters' predictions, I expected to live into old age, so execution at thirty-five was not in the plan. Also, my sister and her family would be very unhappy, and so would Rabbi Mordechai and Master Li. "Well," I said loudly, "I believe I have exhausted all my options.

I can't see a way out, but nevertheless, something must turn up." What could turn up in a tiny cell without a window, lit by a single bare light bulb on a very high ceiling, with a heavy metal door that was barred and locked from the outside, and not a single piece of furniture, I did not know, but physical reality was not the only one I was familiar with.

I decided to try something I had never dared to try before – contacting the Anunnaki themselves.

I had met with some of them, but always through my own masters, or other teachers. Still, I always knew that some day I would have to try it on my own, and this seemed to be the right time. Without debating the issue any further, I acted on my decision. I cleared my mind, ignored my pain and my desire to escape, and aligned myself with my Conduit. If I succeeded, I knew that something, or someone, will indeed turn up.

Almost instantly, a bright yellow shaft of light appeared in the cell. Dusty and shifting, with particles moving in it at random motion, it looked like a very large sunbeam coming through a window in the late afternoon.

The small particles swirled wildly, then suddenly stopped, and coagulated into a globe in the center of the shaft of light, while the rest of it remained bright and empty.
I smiled at it, knowing very well what will happen next, but was entranced by the sight, as always. The speed of its arrival made me suspect that the whole thing was a test – did the Anunnaki, or my masters, felt that I had postponed making contact for too long?

242

They would not hesitate to put me through the horrible ordeal if they felt that this was the only way to make me try the contact. No, they would not hesitate, despite the agony and trauma it would cause me; nor did I resent it, if it were a true test. Sometimes, harsh measures must be taken, as any Ulema would know. I forgot the issue in my joy of watching the light.

The globe burst into fireworks, which then rearranged themselves into the shape of a baby. The baby did not stay small, but started expanding, filling out the shaft in a distorted shape of a human being, whose face was blurred. Very quickly, the shape corrected itself into a proper human form, and stepped out of the shaft of light. Yes, I was honored by the presence of an Anunnaki who came to my rescue.

The Anunnaki was dressed in the long white robe, usually worn by these visitors whenever they came to Earth, and a head covering that hid most of the face, except the eyes, which glittered like those of a wolf. Slowly, the glow subsided, and the eyes became dark, almost black, and unusually large.

I knew these eyes. They had haunted me for ten years, ever since I met their owner when I attended her lecture in Lebanon, just before I was permitted to study the *Book of Rama Dosh*. These dark eyes never left my thoughts, waking or dreaming. I rose from the floor with difficulty, due to the severe pain, and bowed deeply. "Sinhar Baalshamroot. You have come to help me, and I am honored and grateful. I did not presume to think that you had remembered of my existence."

"Of course I remember your existence," she said simply. "I am your Watcher, Germain." I gasped with astonishment and disbelief. My Watcher? Baalshamroot? "I did not know," I said humbly. "I had no idea..."

Theoretically, I knew that every Ulema had his or her Watcher, and I suspected I must have one, but why Baalshamroot? Of all Anunnaki, how did she become my Watcher?

"I chose to be your Watcher after we met at the lecture in Lebanon," she said, answering my unasked question. "When I met you, I knew that you might need me."

Need her? I would certainly agree to that. Meeting Baalshamroot terminated any possibility of my having a normal relationship with an earth woman. This magnificent Anunnaki, so much above me in every way, was still the only one I could ever love, for the rest of my life, or even beyond this life. But of

243

course I could not tell her so; a human is a lowly creature to an Anunnaki, an insignificant creature... and yet she chose me. It was sufficient that she was my Watcher, and it brought me more happiness than a million years with an earth woman could bring. I sighed and asked the question that was important for me to know.

"Was being captured and beaten a test, Sinhar Baalshamroot? Was I supposed to make a contact with you before and neglected to do so in my stupidity?"

"Yes, it was a test. A harsh one, and I am sorry about that, but for whatever reason you kept postponing making a contact, which is a very important skill and we must be sure you know how to do it. I hope you do not resent the serious discomfort this test had brought you."

"I am more than happy about the test. Any amount of pain is worth it, if it makes me know that you are my Watcher, Sinhar Baalshamroot."

"That is good. Well, before we go on, let's take care of the pain from the beatings," said Baalshamroot, and made a motion with her hands. A soft breeze enveloped me; I could almost see a blue tinge in the air around me as the breeze blew and wafted around my body. In a few minutes, not only all the pain and bruises were gone, but the dry blood on my face and hands disappeared, I felt as clean as if I had taken a good bath, and my clothes, which a few minutes ago were filthy and torn, became spotless and mended.

"That is better," said Baalshamroot. "And now, we should leave this miserable place and go on a journey."

"Where are we going?" I asked.

"To the little town of Maalula, near Damascus," said Baalshamroot. "I am sure you have heard of it."

"Yes, I have heard of it, since I lived in Damascus during my childhood," I said. "But I never visited the place. It is a Christian outpost, isn't it, a rather unusual place in a Muslim country?"

"Yes, it is," said Baalshamroot. "But the reason I want us to go there is that it is also my own birthplace."

"Your birthplace, Sinhar Baalshamroot? But you are an Anunnaki. Were you not born on Nibiru?"

"I am an Anunnaki by genetics, my entire DNA is Anunnaki. As you know, some people are, particularly in that part of the world, since it was an Anunnaki outpost long before it was a human habitat.

But like the others of my kind, I was born on Earth, and had to pass through a ba'ab and change my body considerably when I was a very young earth woman, about two thousand years ago. I really was an earth woman, Germain, which is, partially, the reason why I chose to be your Watcher. I do understand earth people very well."

Thoughts were racing through my mind. I heard about a few people who had passed through the ba'abs and became full Anunnaki. I never knew who and why. Could I achieve that? Were my genes sufficiently of Anunnaki origin?

I had no idea. And if I could, and changed myself enough to be accepted by the Anunnaki society, would I be worthy of their respect, enough so as to be able to tell Baalshamroot how much I loved her? By Anunnaki standards, Baalshamroot was very young, she still was a part of our earthly history.
I have heard of earth people marrying Anunnaki who were hundreds of thousands of years old... but I forced myself to stop thinking on these lines. It was not the time to think about such matters; I had to remember, always remember, that I was a lowly creature, a worm, standing next to this glorious, enlightened being.

"Yes, I want you to see my birthplace and to learn who I am, since I will be your Watcher for the rest of your life. And after that, when you pass on to the next phase, we shall remain friends, Germain, perhaps I will even be permitted to guide you further on your evolution. An Anunnaki does not offer such a friendship lightly, nor does she ever desert her charges."

"Thank you, Sinhar Baalshamroot. I have no words to tell you, you cannot know how happy that makes me," I said.

"But I do know," she said simply and kindly. "I do, and it makes me happy, too. And now, let's go." She took my hand, and in an instant we were out of the cell, in the dusty street in front of the prison. Two of may captors were leaning against the prison's door, smoking cigarettes and chatting; they did not see us. Baalshamroot looked at them dispassionately, but with obvious distaste.

"They are evil," she said. "Stupid and evil and unnecessary. Should I kill them? Would you like me to destroy them?"

"No, they are not worth the trouble, they are not the masterminds," I said. "If you kill them, their bosses will find new ones and corrupt their minds, too. We might as well just leave. They may even be killed by their associates, because I disappeared from the cell, anyway."

"Very well," she said. "You are right. Let's go."

*** *** ***

In a blink of an eye, we were in Maalula. The little town itself, which I knew contained less than two thousand people, was not very remarkable. It consisted of small and drab stone buildings with flat roofs, painted tan and blue, huddling against each other in very close proximity.

The air was hot and the light very strong, but it was far from a desert-like environment, since a belt of greenery, including fig trees and grape vines, enlivened the landscape, and patches of herbs and greenery appeared here and there. A sleepy and uninteresting town, but the huge mass of rocks that encircled these houses was more than impressive.

The rocks, situated on the eastern slopes of the Al Kalamun mountains, formed sheer cliffs that dwarfed the human habitation into a beehive perching on the top of an abyss. The slopes were covered with boulders and had deep caverns, some natural, some carved by humans since time immemorial. Baalshamroot looked around her affectionately.

"I have come to visit often since I had turned into a full-fledged Anunnaki. It had not changed all that much over the centuries," she said. "The monasteries are attractive. There are two; one is called Mar Sarkis, or Saint Sergius, and the other is Convent of Mar Tekla, of the orthodox faith. We are going there."

"Why this one?" I said.

"Because it was named after me. I would like you to see it, and a special little cave behind it, when I tell you my story."

"Named after you, Sinhar Baalshamroot?" I asked, perplexed. "But it's called Saint Tekla's!"

"Yes, of course," said Baalshamroot, smiling. "You see, Germain, I was Tekla. In a way, I still am Tekla, though I do not like to think of myself in the ridiculous mode of a Christian Saint..."

I was quite confused, but kept silent and waited to see what will happen, not wishing to annoy Baalshamroot with too many questions, a habit that I once had and took many years to overcome. We proceeded toward our destination. The building was constructed on several levels, giving it a certain elegance, all built from ancient mellow-toned stones.

Following the stairs, we reached the top floor where I saw a church with a dome and a cave into which filtered a stream of water. The whole place was almost empty of people, and had a serene and quiet atmosphere.

"Do you know, the people here still speak Aramaic?" said Baalshamroot.

"No, I did not know that. I knew that it was an enclave of Christianity, but I did not know they spoke Aramaic. I imagine you mean mostly Syriac, the modern Aramaic?"

"Yes, unfortunately, mostly the modern dialect, but it is not all that different. I am sure if Paul preached today, he would have been understood by everyone."

"Paul?"

"Yes, Paul of Tarsus. The one who used poor Jesus of Nazareth as an excuse to convert people to his own new religion."

"So you knew Paul personally?"

"Yes, I knew him well. Very well. Not at first, though. He was already a wandering preacher, well traveled, when I was only eighteen, the daughter of Onesiphorus and Theodosia, a wealthy couple, highly respected in the community.

I was considered a very beautiful girl, and happily engaged to a young man named Thamyris, who also came from a very good family. My parents and his arranged the match, as was the custom then, but he loved me very much. I thought I loved him, and I believed he was so good and kind.

Ah, well, few humans are, really, but at the time I could not tell, and I thought I was human myself. In those days, we knew nothing of DNA, and I have never so much as heard of the Anunnaki. The plans for a sumptuous wedding were being made, when Paul of Tarsus came to Maalula. It so happened that I was sitting at a window from which you could hear every word Paul spoke at a nearby house. His doctrine, during this first lecture, hypnotized me. I thought Heavens opened before me."

"What did he talk about?"

"At first, of Jesus Christ, claiming that he was our savior and the son of God. I was an ignorant girl, since girls were only taught the domestic arts and a little reading and writing. I knew nothing of any spiritual or intellectual matters, but the Anunnaki genetics were strong, and I longed for this dimension in my life, not knowing what I longed for.

Paul's discussion of things beyond everyday reality strongly appealed to me. But what he aimed at, mostly, was the concept of chastity, and that appealed to me even more. He claimed that our views were all wrong. We were taught that our entire existence and our chance to immortality depended on having as many children as possible, and that procreation was a woman's only choice for a good life.

I believed it, of course, and planned on having many children. But Paul was saying that having children, marital relations, any sexual contact at all was bad, wrong, and would prevent our resurrection and our entrance into the Kingdom of Heaven.

I suppose that as a genetic Anunnaki, I already felt that sexual relations, as conducted by humans, were wrong. As you know, Anunnaki couples are united by the Mingling of Lights, not by physical contact, and that human sexuality is merely a faint imitation of the joy of the Union…

I did not know it consciously, but the concept that something else was more important than our idea of the necessity of procreation appealed to me. I listened to him, on and off, for three days, and became convinced that everything he said was true, including his theory about the godhood of Jesus Christ, which later I found to be a pure invention.

I was so convinced of his truth, that I broke my engagement to Thymaris. And that was the beginning of my real troubles, since he was rather vengeful and even cruel. In addition, my parents were distraught."

"So what happened next?"

"If you come into the cave, not this one where the water is dribbling, but my own cave, where I hid later, I will use a Miraya to show you how the events occurred. It is more private than this cave, since pilgrims think this one, with the water, is my cave, and believe the water has miraculous healing properties. We will not be disturbed in the other cave. Some of these events I want to tell you about are hard to believe, unless seen in person.

Mind you, the legends told about me now, thousands of years after the fact, are not all true, of course, and Paul tried to erase my memory from his writings, as well, and with some success."

We climbed a narrow flight of stairs, carved into the wall of the mountain, and eventually entered a small cave. Several large, smooth stones were scattered on the ground. "These stones were here two thousand years ago," said Baalshamroot. "I sat on one of them when I was resting, after I ran away from my pursuers... do sit down, and I'll start showing you the events."

I sat on a large, smooth stone, rested my head on the wall, and waited. It was dark inside after the blinding sunlight outside, and slightly dank, but cool and not unpleasant, and even though Baalshamroot's blue breeze cured me of the pain I suffered from the beating, I was still a little tired, and the respite was welcome.

Baalshamroot removed a small Miraya from around her neck, and directed it toward the wall of the cave. As usual when using a Miraya, a small window of light appeared on the wall, then grew to the size of a large television screen. A picture appeared. A young woman of superb beauty was standing before what appeared to be a court of law.

She wore a simple white dress, lightly embroidered in blue, and a large necklace of gold and turquoise hung around her neck. The girl looked like Baalshamroot, but not entirely. If I had not known these two were the same person, I would have thought they were relatives.

Baalshamroot's face was sculptured and spiritual. Tekla's face was still rounded and had on it the innocence of youth. But you could not mistake the huge dark eyes. Also, the coal-black hair, long and loose over her shoulders, and the clear, glowing olive skin, betrayed the Anunnaki genetics.

The magistrate was looking at her in a severe way. "Why have you turned away from your marriage, Tekla?" asked the magistrate. "Don't you know this is a crime? Are you not ashamed?"

"I wish to follow Jesus Christ and Paul of Tarsus," said the girl. "I wish to learn and to preach the truth of God."

"Paul of Tarsus is a criminal. He was already thrown into prison, and will eventually be put to death for perverting this town," said the magistrate. "I am surprised he was allowed to pass through so many towns and was not arrested before. As for Jesus Christ, he is nothing but a bad dream, my girl. Such a thing as the Son of God does not exist. It is evil and stupid. Forget all that, return to your family, and behave sensibly, or else, I will condemn you."

"I cannot turn back on God," said Tekla.

"If condemned for such a crime, you will burn at the stake, as a witch. Is this nonsense worth dying for? Look at your mother, crying. Look at your father, ashamed of his own daughter."

Tekla did not answer. She just stood there, saying nothing at all, looking at the crowd.

"Burn the witch!" someone shouted, and a whole lot of people took up the chant and cried, "Burn the witch!" Tekla still said nothing.

"Take her home," said the magistrate wearily to her parents. "She is just a stupid and misguided child and I have known your family for many years. Put some sense into her head. Beat her, if necessary. However, if she continues with this, she will burn. I am warning you."

he Miraya darkened, but something started whirling, and I realized Baalshamroot was going to show me the next turn of events. Indeed, I saw Tekla running in the streets and approaching a dismal house. She knocked on the door and a man, wearing a shabby Roman outfit, came out. I assumed he was a turnkey.

"Let me in, and I'll give you my earrings," whispered Tekla. "They are made of gold." The turnkey looked around the street, saw no one, and let her in. She handed him the heavy gold earrings, and he put them inside his robe, quickly.

They walked to a cell, and he let her in there, too. Inside was a tall, thin man, sitting on the straw that was spread on the filthy floor.

Tekla threw herself at his feet and cried. He began talking to her, telling her to be strong in her faith. It was horrible to watch, since I knew that he was going to lead the silly girl to her death as a witch, and there was nothing I could do about it. I wanted to kill him. I had to remind myself that the event took place two thousand years ago, that the same girl was standing by my side, but I could hardly bear it anyway. Two men entered, grabbed Tekla off the ground, and led her away.

"One of them, the good looking young man on the right, is the man I was engaged to," said Baalshamroot calmly. "He followed me, and his brother came with him. They all felt their family was disgraced because I broke the engagement."

"The magistrate told your parents to take you home..."

250

"Yes, you see, my parents did take me home and locked me up in my room, and after I refused to listen to them they beat me, but when they left me finally I climbed out of the window and ran out to find Paul, and the scoundrel kept inciting me to continue with my idiocy. Well, here is the next scene, watch."

I saw that the two men brought her in front of the magistrate again. This time the whole thing took no time at all, and the magistrate said, "I give up. Burn her."

I saw people stack the wood, build it high, and prepare the stake. They tied her to the stake and lit the fire. I saw Tekla looking around wildly. "Who were you looking for?" I asked.

"Jesus Christ," she said. "Can you believe it? I thought he would come to me. But look, someone else came."

A great eruption from the earth was heard, from some distance. "My Watcher," said Baalshamroot, smiling. "Her spaceship made a noise, since she was in a great hurry to get me out of the fire and executed a bad landing." Light was pouring suddenly out of the sky, a cloud formed, and huge quantities of rain and hail came out of the sky and quenched the fire.

"And I still thought it was Jesus Christ," said Baalshamroot, laughing at the memory. "The crowd ran away in terror, and the magistrate let me go. They all panicked." I saw people undoing the ties and telling her to leave town. The Miraya darkened again.

"I left my parents and my town. It so happened that Paul managed to run away from jail, I don't remember how, and he hid in a cave with two of his people. I joined him and begged him to let me come and be his disciple.

I must admit that for a while he tried to dissuade me, telling me that women always get in trouble when they preach, particularly pretty ones. But I said to him that I was willing to take the chance, and anyway, I no longer had a home. So finally he agreed, and we went to Rome.

It was a long journey. The legends later told that I went to Antioch, where I became the first Christian Saint, but this is not true. We passed Antioch on the way, but the events that they are telling about, at the arena, happened at the Coliseum in Rome. We preached and carried on with sermons and baptisms, converting many poor people into Christianity.

I believed with all my heart that Jesus Christ was my savior... and all the while the real Jesus of Nazareth was living comfortably with his wife, Mary Magdalene, and their children, in Marseille. He went there after he was saved so cleverly from the Romans, but that is another story.

Ah, well. I was young and innocent. Thank goodness, my Watcher was still taking care of me. Yes, I was arrested again, they caught me preaching. Paul was out of town when it happened, converting and baptizing in the rural areas next to Rome. After a short hearing, I was taken to the arena, to be thrown to the lions. Look." She directed the Miraya again.

The Coliseum, clean and new, was packed. The Emperor and Empress, beautifully dressed for the festivities and hailed by the crowd, came in and sat down under their velvet and silk canopy, ready for the fun of seeing human beings torn to pieces by beasts who were already tormented by hunger and thirst. Two victims were thrown into the arena, one man and Tekla.

The crowds cheered with joyful expectation of the blood sport. Heavy iron gates were raised up, creaking, to release the great beasts. Two male lions and a lioness entered and looked around, dazed by the light.

They were thin, probably starved for days, but still majestic, their tawny skin glowing in the afternoon sunshine. The male lions wandered around the arena, somewhat confused, shaking their impressive manes.

The lioness, on the other hand, marched straight to Tekla, and stood before her, as if she were a faithful dog. It was strange, she did not attempt to get a meal, just waited patiently.

The male victim was quickly killed by a male lion, who calmly began to eat his flesh. The crowd, slightly disappointed by the speed, cheered a little feebly. They would have preferred to see some torture inflicted on the victim by the lion.

The other male lion came forward, crouched, and attempted to leap over the lioness to grab Tekla. Very neatly, the lioness leapt into the air and with one bite on his neck, killed the male lion, who was much bigger than her.

The crowd screamed. The second lion raised his huge head, noticed the commotion, and deserting his meal, tried to attack Tekla over the body of the lioness, exactly as the first one did. The lioness jumped against him and bit him, too, but this time she was already tired and the bite did not kill him outright, so the two beasts engaged in a fierce fight, goring each other, and eventually dying together.

The starving lioness, never attempting to get her own meal, sacrificed her life to defend Tekla.

Such a thing could not have happened, and yet it did. It must have seemed entirely unnatural to the crowd. The young woman, alone in the arena, surrounded by the remains of the other victim and three dead lions, stood tall and steady, gazing at the crowd with untroubled, fearless eyes. She even smiled at them.

Several women screamed, "Free the girl! The gods are protecting her!" Tekla heard them and shouted something, but could not be heard over the crowd's screams. "Would you believe, I was saying that there were no gods, that Jesus Christ himself was protecting me," said Baalshamroot. "No one heard, though, which was a fortunate event. Look."

Many women joined in. "Free her, free her!" they were screaming. The Emperor looked rather helpless, not knowing if he should unleash his guards on the crowd, or obey it. The empress got up. "She is to be free," she commanded, and lowered her thumb in the traditional gesture.

"The woman has subdued the wild beast. She must be a virgin, protected by Vesta. I command her release." The Emperor seemed happy that the decision was made for him, and lowered his own thumb. A slave went into the arena and lead Tekla out.

"The Empress held her thumb down, Sinhar Baalshamroot," I interrupted, surprised. "Shouldn't it be up? She gave you freedom."

"No, this was a mistake, perpetuated in Hollywood, where they did not read Latin very well. It is exactly the other way around."

"And why did the lioness defend you?"

"My Watcher, of course, controlled the lioness," Baalshamroot said. "I left town and went in search of Paul, preparing to continue our mission. I met him out of town, and we went on, but everything changed between us. After the scene with the lions, many people came to believe I was a miracle worker, and my reputation preceded me. People knew about me before I came to various towns, and seemed to pay more attention to me than to Paul.

He did not like it. Paul did not mind having me around when I was a humble follower, but generally he disliked and mistrusted women, as is clear from his writing. When I became famous, he was afraid I would usurp his power and control over the events. Therefore, he betrayed me.

It was easy enough for him to do so when I tried to visit my parents in Maalula. I heard that my vengeful betrothed, Thamyris, was killed in an accident.

He had been drinking heavily with his friends, and on their way home, a wagon, pulled by an ox, ran over him. My father was also dead, and I wanted to be reconciled with my mother."

The Miraya showed the little town, not all that different from the modern one. Tekla stood talking to her mother. "But Thamyris is dead, Mother," she said. "He will no longer claim me as his wife. Why can't you forgive me? Why can't you join me in my belief in Jesus Christ, our Savior?"

"Thamyris' death does not make you less of a criminal," said Tekla's mother. "He would not have been out that night, drinking, if you had married him. As for your father, he died of shame. It is as if you killed both, and ruined my life as well. What have I got to live for? Go away, you brought me nothing but pain and shame."

Tekla lowered her head and went away. The Miraya followed her to a water pit, surrounded by many people. Paul stood by. One young man suddenly said – look, everyone! This is Tekla! She is the one Thamyris died for! Catch her, kill her!"

I saw Paul slink away behind some people, as a group of men caught Tekla and held her. "What shall we do with her?" One asked.

"Throw her into the pond! Drown her! Let her die for what she had done to Thamyris, and her father, too!" shouted another man. "Drown her, kill her!"

Baalshamroot sighed and stopped the Miraya for a minute. "My poor Watcher must have been getting tired of protecting me. So many times she had to extricate me from all the entanglements.

But this time, I saw Paul run away and leave me to my fate, and for the first time, my faith in him wavered. I was so shocked to see him betray me, not coming to my aid, that I felt my entire world was tumbling down around me. My believe in Jesus Christ was closely associated with my belief in Paul's goodness. If Paul betrayed me, was Jesus Christ true to me? In an instant, I lost my faith, lost everything.

I wanted to die, death was the only release from my agony. With superhuman strength I extricated myself from the men who held me, and threw myself into the pond. Look."

I saw Tekla throw herself into the water, but she did not sink. Instead, she rose to the surface as if weightless, her long black hair floating like a cloud around her. She seemed to be shocked, looking around her with dismay and horror, since she obviously did not try to swim, and yet she was not drowning. A great yellow light appeared in the sky, shining right over her, and she was slowly raised into the air, still in the same horizontal position she was in the water, her wet hair tumbling down vertically and dripping water.

The crowd screamed with terror. "She is a witch! Run away! She is not drowning!" they all turned and ran away, in panic, as she was levitating in the air toward the light. She floated in the air for some time, then seemed to turn as she was set on her feet on the edge of the water.

Baalshamroot turned the Miraya off. "As I was set on my feet, a gentle voice said in my ear, "Run to the cave above you, Tekla."

I ran up to the cave, obeying the voice mindlessly. I went inside and sat on one of the stones, dripping water all around me. What was happening? I had no idea. Outside, I could hear the people, those who recovered themselves from the panic, climbing the mountain after me. That was fine with me, let them kill me, I thought. I want to die anyway. What did I have to live for? But it was not to be.

As the people gazed into the cave, hesitating weather to come in and grab me or not, a shaft of light appeared in the cave. The people retreated, scared of the sight. Out of the shaft came my Watcher, took my hand, and guided me a little deeper into the cave, where her spaceship stood, ready to leave. We entered it, and she took me away.

At the time I did not know what happened to the people, but later I heard they had told everyone that I went into a cave and then vanished into thin air."

"And after that, did you understand the situation? Did you realize who the Watcher was?"

"Yes. Once I got over my ridiculous faith in Paul of Tarsus, I could begin to understand the truth. From then on, my life changed, and I worked toward becoming the Anunnaki I am today, but that is another story, for another time; some day I will tell you how it was accomplished.

The interesting thing is, Paul had tried his best to erase my name, and greatly succeeded, at least for a while. My name only appears officially in a small book, called the *Acts of Paul*. But the legends that begun to surround me were impossible to kill. It is still claimed that I was the first Christian Martyr, that I have sacrificed my life for the Son of God. Thank goodness this did not happen, and I can devote my life to learning and truth."

"An amazing story. Thank you for telling it to me, Sinhar Baalshamroot."

"It had to be told. And now, I will be taking you home to Paris. Or would you prefer to visit Rabbi Mordechai in Budapest?"

"Yes, I would love to see him, tell him all that had happened."

"He knows, Germain."

"So that is why I could not contact anyone telepathically? He blocked it for the test?"

"Yes."

"And I passed the test?"

"Yes, Germain. You passed the test."

"Will I see you again, Sinhar Baalshamroot?"

"Yes, now that we have established contact, I will be visiting you off and on. And of course, should you need help, please contact me. Always remember I am your Watcher."

As if I could ever forget..."Thank you, Sinhar Baalshamroot," I said.

"Will you report to the Pères du Triangle soon?"

"As soon as I am back in Paris, after my visit with Rabbi Mordechai."

"And do you feel ready for your next mission?"

"I am ready, Sinhar Baalshamroot. Always. The Pères du Triangle know it." She smiled at me, pleased that the ordeal did not deter me from my work. I smiled back, happy in the certainty that she will not disappear from my life. I would never feel alone again, knowing that she was there for me, my own Watcher, my guide forever.

In a blink of an eye, I found myself in front of the familiar house of Rabbi Mordechai in Budapest. He was standing in one of the windows, waiting for me. I waved at him and walked straight in.

119. Jesus Did Not Die On The Cross
⌘ ⌘ ⌘

Ambar Anati is telling us how she went back in time, and visited Mary Magdalene and saw Jesus Christ in the flesh, living and working in Marseille, France.

Story told by Ulema Maximillien de Lafayette and Dr. Ilil Arbel:
Here is her personal account, in her own words, verbatim, and unedited: "My first shape shifting and time travel; how I met Jesus and Mary Magdalene – who turned out very differently than expected.

It is impossible to describe my feelings regarding my expected visit with Mary Magdalene. I always had a deep interest in her and thought she must have been a most complicated personality, and her place in history, next to Jesus himself, was enough to make me revere her.

The thought of seeing her, and soon, was an emotional whirlwind, particularly since Sinhar Inannaschamra was quite sure that I will be able to meet Jesus as well.

Having been brought up Christian, it was difficult for me to accept the possibility that Jesus not only did not die on the Cross, but lived happily in Massilia (modern Marseille) with his wife and family. But the Anunnaki have never been wrong in anything they had ever told me, so I decided to put my disbelief and shock on hold, keep an open mind, and learn as much as I could.

I assumed I would have to use a cone to learn the language spoken in ancient Gallia Narbonensis, the part of France where Massilia existed, prepare my clothing, and practice shape shifting, but it turned out I was wrong.

All I had to do, said Sinhar Inannaschamra, was to wear my Monitor around my neck, exactly the way I wore it when I went to my mission at the Grays' base, and use the same code, so she could keep an eye on me. She was going to check on me periodically, and if anything went wrong, come and get me. But she expected everything to go smoothly, and as always, I trusted her. She gave me the exact directions as to what to do and how to behave, and indeed, it seemed so simple I could hardly believe it.

"How do I get back?" I asked.

"Very easy," said Sinhar Inannaschamra. "Just speak the code to the Monitor and think of Nibiru. It will do the rest, you will hardly feel it. This trip should be very comfortable, it's just training, you know."

Next morning I went to the Akashic Library, wearing normal clothes and carrying nothing other than the Monitor around my neck. I repeated the steps I took when I met with the Council, only this time requested the Departures Room. The pad took me to an octagon-shaped glass pavilion that stood out of the main building, surrounded by ancient conifers.

The glass was transparent, so I could see that the room was entirely empty, other than one stool in the middle of it. The room was not very big, and it had a glass door, also transparent. I entered, and to my amazement, I could not see anything outside, because the glass became opaque.

I had no idea at the time what caused it, since Sinhar Inannaschamra forgot to tell me about it. Later I found out that the air in the room has a special quality that made the glass opaque.

As instructed, I sat on the stool, and that started the action. One of the panels opened up, and millions of charts with codes started to flicker on a screen. At the same time, all the other panels turned black. The whole room darkened, and only the front panel, which showed the charts, had any light on it.

At this moment, the screen seemed to release a plasma-like substance, and it came toward me and circled my body. I began to spin around myself, feeling very dizzy and strangely heavy. I am not sure how long it lasted, but it seemed that in an instant I found myself in an ancient town by the sea.

To me it seemed not much larger than a fishing village, but this was only because I was not used to first century towns and had the prejudices of a modern person as to what a town should look like.

This was Massilia, the Roman name for present day Marseille, a thriving, growing Roman trading port, and I knew it did well because it was the first town of Gallia Narbonensis to have a public sewer system. Actually, it was already an old town, since originally it was built by the Greeks and had archaeological remnants of the Greek settlement. But the Romans had added quite a lot of buildings, streets, and roads since Julius Caesar conquered it.

I knew what to expect, but the experience was nevertheless incredibly bizarre. Most people seem to think that time travel is simply going to the time and place you wish to visit, and entering it as if it were a normal place. This is not so. When time traveling, it is almost as if one travels in virtual reality, and the experience has to build upon itself, to materialize. Therefore, the place was entirely empty, with only a few fishermen standing at a distance on the shore. But they were not moving, they were like statues.

The rest of the town was completely empty of people. I felt the town was alive, as if expecting something to happen, but no one was in sight. All of a sudden, people started to appear in the street, but they came from nothingness, from the empty space in the air. Materializing one by one, they filled the town; it felt as if I was meeting apparitions. Then, in a few minutes, I adapted, and started experiencing real life in a real town. Everything became normal.

I had to take a short walk just to study the town a little. Massilia was a bustling town. The houses were made of stone fixed with mud, low and sturdy, but a few houses were truly elegant.

Walking toward the opposite direction from the sea, I entered the open market, full of little stores and workshops. Each workshop was a small, rather dark room, that opened on the street with a large, open doorway.

You could see the craftsman working on the goods he hoped so sell, or the merchant surrounded by his goods. Most of the stores were devoted to foodstuffs, such as stores of neatly packaged spices, or one devoted to cheeses and pickles.

But there were also some shops that sold fabrics and various notions needed for weaving, sewing, and embroidery.

The market, with its noisy cries of the merchants, the interesting smells, and colorful sights, was extremely appealing. After enjoying my visit with the market, I turned back and walked to the shore. The whole place smelled like fresh fish, but it was not terribly unpleasant despite my growing dislike to animal food. Those fishermen who were not at sea were mending nets, cleaning their catch, and doing other chores. Some little fishing boats were turned upside down, being mended. The sun was shining, the light Mediterranean breeze was blowing, and beautiful, big white shells were strewn everywhere. I picked a little warm sand in my hand.

Having been brought up in Maine, I was very fond of the sea. The place felt like home and I felt a twinge of homesickness, remembering how my mother would bring me to the shore and help me collect shells and smooth sea glass and pebbles.

I decided the time came to drag myself away from nostalgia and go visit Mary Magdalene, fully aware that I was postponing this very thing I wished to do because I was a little nervous about it.

Resolutely, I turned away and went into the town proper. Many women were carrying baskets on their way to the market, and to my relief, as I looked at myself, I was assured that I was wearing the same clothes as any of the better dressed women who walked the streets.

I had on a long woollen outfit, white with multicolor stripes on the sleeves and on the hem, soft leather sandals, and a thin silk scarf loosely covering my head. Obviously, I matched the upper class ladies, just as Sinhar Inannaschamra told me I would. I took my Monitor, which was still hanging around my neck, and looked at my face in the mirror that was part of the Monitor.

I looked different than the usual, in a subtle way. The spinning that turns into shape changing, really affects all the molecules in one's body, even helps one fit into the climate and environment one travels to.

Sometimes one does look different for the duration of the trip. I looked a little older, I thought. That was good, I would inspire confidence in Mary Magdalene as an older woman. I must say I liked the ornate silver earrings I was wearing, just showing under the white silk scarf.

Very nice, I should get something like that when I was back on Nibiru, I thought. The only thing left was to rehearse the language. At the moment of arrival, I did not know the language, but all it took was hearing one or two words spoken on the street. This is because the spinning opens up the Conduit, and triggers the part of the knowledge depot which is the seat of all languages.

As soon as I heard those words, I knew the language as if I had spoken it all my life. I rehearsed a few words, just in case, and felt comfortable.

But I was determined to speak to Mary Magdalene in Aramaic, her own language. I looked at the Monitor, got directions to Mary Magdalene's house, and strolled there, still enjoying the sights and sounds of the busy town. I did not have far to go to the small, neat house, well maintained, and standing in a little garden planted with herbs and flowers. Like all the other houses, it was made of stone and mud, and two stories high.

I stood on the other side of the street, wondering how to approach Mary Magdalene. I was so excited about the meeting, that I did not think about what lines to use to persuade her to talk to me. But everything worked out very well.

As luck would have it, I saw a woman approaching the house, carrying a basket. One look and I knew it was Mary Magdalene, because Sinhar Inannaschamra showed her to me on the monitor. She was not beautiful, exactly, but nevertheless was extremely attractive.

She must have been in her late thirties, small and slight but with an elegant figure, and she carried herself with poise and dignity. She had lovely, warm brown eyes, and dark chestnut hair slightly touched with white at the temple, which I could see because her scarf was sitting way back on her head. She was simply and neatly dressed in an outfit very much like mine.

I approached her, and greeted her in Aramaic, introducing myself by the name Ambar-Anati, which I was sure would be more familiar to her than Victoria. She seemed extremely surprised and delighted. "Are you from Judea, Ambar-Anati?" she asked in a very pleasant voice.

"No, I am Phoenician," I said. A half truth, but I really could not do any better.

"It is nice to hear you speak Aramaic," she said. "My husband and I speak it at home, and taught it to the children, but most of our friends here do not know it. Won't you come in and rest, and have some refreshments? And tell me what is it that you wished to speak to me about?"

I accepted with pleasure, and we entered the living room, a simply furnished but very clean and pleasant place. She went to place her basket elsewhere, probably a separate little room devoted to cooking, thus giving me time to observe the room carefully.

This was clearly the house of a middle class family. It had many comforts and conveniences, though certainly not ostentation or overt luxury. The walls were neatly plastered and whitewashed, the floor was tiled, and the windows had a lattice structure that provided security and decoration at the same time. There were three or four niches in the walls, each containing an oil lamp.

This family obviously did not go to bed with sun, as poor people were forced to do; the lamps spoke of reading and writing and spending time with family and friends after the sun had set. A few small rugs covered the floor.

The room was very adequately furnished with a large table with two benches, each with a few colorful, embroidered pillows on it, storage boxes made from beautiful dark wood, and a built-in stone pallet that had a throw and pillows providing comfortable sitting. Here and there, on the rugs, there were also large cassock-like pillows covered with beautiful fabrics.

A couple of copper braziers stood by the wall, awaiting the season of winter and glowing softly. In a corner stood another large table, covered with manuscripts, including one that was currently worked on, writing implements, and inks.

The whole room was scented by a big bunch of cut herbs and flowers in a clay jar, that stood on the dining table.

Mary Magdalene came back with cold water, wine, cakes, dried figs, and honey.

"As you might know, the Romans decided to make a law here that women cannot drink wine when a man is not around, but we don't pay much attention to it... they don't really follow you into the privacy of your home, I must say." She poured me a glass of wine. The wine was delicious, and she pressed some of the food on me.

"I noticed that your beautiful throw on the couch is white and striped with blue," I said. "These are our colors in Phoenicia, the symbolic colors of the god Melkart."

"Well, there is little difference between our nations, and white and blue are our colors as well. I knitted this throw, thinking all the time of Judea..." said Mary Magdalene. "Our nations are related, you know.

Not only through the marriage of King Solomon and King Hiram's daughter, but even before. I like to hear Yeshua tell me, and the children, about the history of our people. And he told us that many people believe that Joshua, the one that helped Moses during the Exodus, really was a Phoenician Prince.

As a matter of fact, he entered Canaan independently, from the north, and settled peacefully. He never even knew Moses, they say."

"I had no idea," I said, making a mental note that this would be a subject worth pursuing at the Akashic Library on my next visit there. "This is fascinating. I really must look into it. I like history too, you know, very much."

"You must visit us often, then, and discuss this matters with Yeshua. He will be happy to meet another enthusiast of his favorite subject."

"I would love that... I promise I will come back, if you will let me. But I might as well tell you what I came to ask you, before I go on enjoying your hospitality," I said, a little guiltily. "You may be angry with me, since I am about to rake up your husband's past."

"I am rarely angry," said Mary Magdalene. "And I don't really mind talking about the past, as long as it is to another woman. I am still afraid of men, though. I always feel we are forever in danger.

I constantly warn my husband to stay out of trouble. Now that he is older, he listens better. Yeshua is a very nice, kind man, and he does listen to me when I advise him on many matters, but sometimes I wonder if he understands that we should be careful for as long as we live. He is very intelligent, and extremely well-educated, but between you and me, he has absolutely no common sense."

"I know little about your husband, only that he was falsely arrested by the Romans and you had to leave Judea."

"Indeed, that is the truth. I always knew we would get in trouble," said Mary Magdalene, her smile disappearing. "As I said before, and I mean it, Yeshua just did not have any sense whatsoever.

He was a healer, and he had great success in curing many people. Unfortunately, he was also a bit of a magician, and instead of keeping his talents to himself, he would go and perform his healings and miracles in front of important people. They hated him."

"But they could not object to his healings?" I asked. "Nothing is wrong with making people feel better."

"Even healing can seem to be blasphemy, particularly if you also do magical tricks as part of your performance. There was even some foolish talk about his making the dead rise – of course this was sheer nonsense, the person that 'rose' from the dead, his name was Lazarus, a relative, simply fainted and Yeshua made him feel better – but such stupid talk would cause trouble.

Yeshua never even heard of the story – he was away, with his friends the Essenes, when it circulated, and I kept it from him when he came back. But his talents of healing got him a bunch of followers, disciples of sort, and they were good for nothing. All they wanted was magic, sensational tricks, and they went about saying blasphemous things about Yeshua being the Messiah. Naturally, the Sanhedrin, once they heard the word Messiah, took a dislike to him."

"They would be sensitive about it, I suppose," I said. "The Messiah is an important issue with your religion."

"Well, yes, and I must admit Yeshua was quite annoying," said Mary Magdalene, smiling somewhat indulgently at the memory. "The business of having disciples made him think himself of more importance than was good for him. He insisted on preaching, and told people he was the Son of God.

Now, that was a common thing to say if you knew the Essenes, a group of desert recluses he once lived with and kept on visiting; actually he spent quite a long time with them, enough to get him to believe in much of their doctrines.

They call every honest person 'Son of God.' But the authorities did not like it. Again, it sounded to them like blasphemy.

The Sanhedrin members were very set in their ways, except for one man, Joseph of Arimathea. That is because Joseph was an Essene, too.

No one knew it, he kept it a secret, since belonging to a sect would have spoiled his business and his reputation, but he never gave up the connection.

And of course there was Nicodemus, his young protégé. He was also an Essene. Joseph and Nicodemus were real friends to us. I don't know what we would have done without Joseph, he really handled everything when the trouble began."

"And Yeshua was tried before the Sanhedrin, right?"

"Not right away. First, he was arrested and taken to an interview with the Procurator."

"Am I right that this was Pontius Pilate?" I asked. "Yes," said Mary Magdalene. "That was him. Do you know he was replaced a few years later?"

"No, I had no idea. Why was he replaced?"

"He was accused of some crimes, which of course he did not commit. He simply fell out of favor. Joseph told us, on one of his visits to Massilia.

He was very much surprised when the replacement happened. The Romans are sometimes very cruel; it is possible, though we are not sure, that they forced him to commit suicide. Anyway, Pontius Pilate was not terribly interested in Yeshua... it seems he even wanted to acquit him.

We know a little because his bodyguard, a Roman centurion, heard everything that was said when Pilate spoke to the representative of the Sanhedrin, and later wrote it all down. That is because he knew Yeshua, who once cured his daughter from a terrible illness.

He really saved her life, and from a distance, too. Yeshua never saw the child; he was good at such things. The Roman centurion was very grateful to Yeshua, and thought it would help if he took those notes for posterity. I imagine he knew that between the Romans and the Sanhedrin, they would execute Yeshua. But there was more to it than just this interview.

As I said, Pontius Pilate probably would have let Yeshua go free, because he could not care less about religious matters, which were the chief complaints of the Sanhedrin. Rather, Pilate asked him if he had any designs against Caesar, and Yeshua answered that of course not, he had no problem with Caesar at all, he knew Caesar ruled Judea legally.

So Pilate asked him if he would admit to Caesar being the strongest god, which was the standard thing to ask, and that was Yeshua's downfall. I would have advised him to admit Caesar's superiority, since it was a private interview and none of his friends was there to hear it.

But Yeshua could not bring himself to blaspheme against God. The fool, all he had to do was just nod his head... and what is more, he went very far in his protestations about Caesar.

He told Pilate that he, Yeshua, was more powerful than Caesar because there was only one God, and he was the son of God. Again, more Essene nonsense. Naturally, after that, Pilate simply had to turn him over to the Sanhedrin, he had no choice. Someone, years later, was circulating the rumour that Pontius Pilate never forgot Yeshua.

Apparently, he was quite interested in his capacity as a healer, and even intended to send him to Caesarea, to be his own healer, since he had some illnesses. Ah, well. Sometimes, the Sanhedrin can be more cruel than the Romans."

"And what happened then?"

"A huge, famous trial took place. The high priest, Caiaphas, was after Yeshua's blood. He was a Sadducee, you know, one of the rich, higher classes. He felt that Yeshua was a threat to the usual order of rich and poor, high and low... you know, a rebel. Caiaphas was a really nasty man, eager for power. And he had power, lots of it...

Our friend, Joseph of Arimathea, did a brilliant job of defending Yeshua, and it would have gone well, but for once Joseph made a horrible mistake. He questioned him, 'Do you consider yourself the Messiah?' and Yeshua denied that in the most sensible manner; after all, Yeshua never thought of himself as the Messiah! Never came into his head to imagine that! Then, Joseph asked the question that destroyed everything.

He said, 'Who are you, then, Yeshua?' and Yeshua was stupid enough to say, 'I am the Son of God.' I know, I know, he should have known better, but he was very foolish at that time. That gave the Sanhedrin the excuse to send him to the crucifixion.

Caiaphas practically jumped with joy when he heard Yeshua say this thing. Horrible, horrible man, Caiaphas. I will hate him as long as I live, and believe me, I am not quick to hate."

"So he really was crucified," I said, sadly.

"Yes, Yeshua was crucified. I cannot tell you how cruel, how horrible, this practice is. He suffered so much, blood all over his body. I was with his mother, and one of his disciples, but nobody else.

All the other disciples ran away, they were scared to death, fearing they would be arrested by the Sanhedrin. It broke my heart that these people, who always claimed to love Yeshua so much, were not there for him as he was dying.

A few people gathered around, probably just curious people, and they stood near the cross, but not very close, because the Roman soldiers did not allow them to do so. His poor mother collapsed twice in my arms."

"His mother, Mary of Nazareth... Yes, please tell me, what was Mary like?"

"Mary... I miss her so much. She was very kind, always so sweet. I loved her very much, she never said an unkind word to me. And she had every reason to be mad at me, because I broke God's law and lived with Yeshua before we were legally married, I am ashamed to say..."

"But you always meant to be married, so it does not signify," I said.

"Oh, indeed, we always meant to be married. It was just because of all these delays and troubles, and things sometimes just happen when you are young... Still, many other women would have held it against me. But not Mary. She was too kind."

"What did she look like?"

"She was incredibly beautiful," said Mary Magdalene, her eyes misty with the memory. "I have never seen anyone as beautiful as Mary. She was only fourteen when Yeshua was born, so she was still young when the trouble happened. She had very long black hair, which she always put in one long braid. Her skin was white, like the finest ivory, and she had big, clear blue eyes, rather unusual for our people.

She always wore a lot of blue, she was a little vein about her eyes and about her great beauty, but not in an unpleasant way, and who could blame her... My daughter Sarah inherited these amazing eyes, every time I look at her I think of Mary..." She wiped a tear.

"Do you know, Mary's hair began to gray very quickly after Yeshua's crucifixion? It only took a couple of months before it was all silvery white; it must have been the agony she went through, seeing her son undergo such pain. But even with the silver hair she looked young and beautiful.

Perhaps even more beautiful. There was something so delicate, so soulful about Mary."

"So the two of you stayed by Yeshua's side. It must have been horrible."

"We just stood there, crying and helpless. The men on the two crosses on Yeshua's sides fainted, off and on, like Yeshua. Then, Nicodemus came and asked the Roman soldier, who was guarding this row of crucifixes, if he can wipe the face of Yeshua, and give him something to drink from a sponge.

The soldier said 'Yes, go ahead,' so Nicodemus dipped the sponge in a bucket and brought it to Yeshua's mouth. I saw him sipping from the sponge, and blood kept pouring and pouring from his hands and his feet. A few moments later, it was clear that he passed away; Mary fainted and fell to the floor.

As for me, I felt this was not happening, as if it was a nightmare, and I was expecting to wake up. As if in a dream, I approached the Roman soldier and asked him if I could take Yeshua home for burial.

I simply could not bear to leave him there on the cross. But he said that this was against the rules. I asked him, 'What are you going to do with his body? He is dead, after all.' The soldier told me that the law requires that all crucified people first be checked to see if they were really dead, because sometimes it takes them two or three days to die.

And after that, the Romans would take them and dump them in a place reserved for crucified people and other condemned dead prisoners."

"This is disgraceful," I said

"Yes, it was very hard. But we were helpless. What could two women do against the Roman soldiers? So we left, and returned to the house, where we kept crying all of the late afternoon and evening.

Suddenly the door opened, and Joseph of Arimathea came in. He looked hurried and upset. 'I have some news for you,' he said. 'I went to Pontius Pilate and I asked him for a favor. I know Pilate through business, so they let me in.

I asked him if I could take Yeshua's body to be buried in my own family grave. Pilate said 'Fine, go ahead.' To tell you the truth, Ambar- Anati, I don't think Pilate cared one bit about anyone in Judea. He was so bored with us and all he wanted was to get out of this horrible job."

"So Joseph got the body? What did he do with it?" I asked.

"I can repeat the exact conversation, I remember it like yesterday," said Mary Magdalene. "Joseph said, 'this is a great secret, which you cannot tell anyone, especially Simon and Peter and the rest of the disciples.'"

"I will say nothing," I said to him. "Just tell me what you have done."

"I had to do something very quickly so no one will find his body," said Joseph. "I got Nicodemus, and we took Yeshua right away to the grave of my family, making sure everyone saw that. Then I put some of his clothes there, arranging them so that they would look as if they contained the body, but we only stayed there for a few minutes."

"Clothes? What did you do with the body?" I asked him, perplexed.

"As soon as we were alone, Nicodemus and I transferred Yeshua to another place, which for the moment must remain secret. We rushed to do it as fast as possible, which was extremely lucky, because as soon as I came back to my family grave I saw Roman centurions marching toward it.

I asked the Romans what did they come for, and one of them told me that they got an order to guard the tomb, and they must seal it first. I did not want to ask them who gave the orders, but I suspect it was some of my friends at the Sanhedrin...

The Romans helped me to roll the big stone that usually sealed the grave. I felt so relieved, since I knew no one will get in anymore, and no one will ever know that Yeshua was not there."

"I don't understand anything, Joseph," I said. "Why did you have to go through all that? Why not simply bury Yeshua properly?"

"Just wait, Mary," he said. "I find it very hard to explain. I went your relative, Elizabeth, and told her what I did. She said that she would like to go and anoint the body in preparation for the burial. So I told her, 'You don't need to do that.' 'What do you mean?' she said. 'This is out tradition!' So I told her the truth. 'Yeshua is not going to be buried. He is going to be all right.' She thought I was crazy, probably you think I am crazy too."

I interrupted him, and I asked, "Are you mad? Are you trying to tell me that Yeshua is alive? How can that be? I saw him die on the cross, right after Nicodemus washed his face and let him sip some water. Did you witness a miracle? Is Yeshua really the son of God?"

"It was not a miracle, Mary, but the water were not plain water, either," he said. "I know something about herbs, from my days with the Essenes. I put a very strong herbal concoction in the water, one that creates a death-like state that would last a few hours. And with the blood that Yeshua lost, and his weakened condition, he will be like dead for at least a night, but then he will wake up."

My mother in law heard all that in total silence, in complete shock. She obviously could not accept the good news so soon after the horrible ordeal. But then she said, "I must go to him, right away. I have to help my child, whether he is dead or alive."

"You cannot do so, my dear," said Joseph gently. "If you don't stay at home, and receive your friends and neighbors' condolences, the Romans will suspect something. Tomorrow morning, very early, Elizabeth and Mary Magdalene will come to see him. But you must be brave and stay here and pretend that Yeshua is dead and buried. It is essential if we are to save his life."

And this is exactly what we did. We stayed all night. The next day, Mary stayed home, but Joseph came and led me to a cave, quite a distance from our village. Nicodemus was guarding the entrance, and moved to let us in. Yeshua was lying on a large stone, which was covered with some soft blankets.

He was exhausted and could not talk at all, but he was alive! Joseph saved him! I immediately saw that Joseph thought of everything. There was a bowl of grapes and olives by Yeshua's side, and some flat bread, and a jar of water. And Joseph brought clean clothes for Yeshua.

We treated his wounds as best we could, cleaned him, and dressed him with the fresh clothes. He felt better and could mumble a few words of thanks, but he was not entirely conscious.

"What will happen now, Joseph?" I asked.

"We are going to take Yeshua away, to the house of two fishermen I usually do business with. They are very loyal to me, and one of them is from Tyre, which will ease our escape."

We went there, a rather long journey, Joseph and Nicodemus carrying Yeshua on a large board which they covered with the blankets. The fishermen were waiting for us. They were really very scared, but they were trustworthy and kept their word.

We took Yeshua in, we entered the house, and there were two women there, one of them a Phoenician woman who rushed to help us, and took Yeshua to a small room where he could rest. We spent a short time with him inside the room, while the two other men were guarding the house.

"Mary," said Joseph, "Yeshua must leave Judea right away. He is in grave danger here."

"That is fine," I said. "I have some money, perhaps we should hire a boat?"

"Do not concern yourself with that," said Joseph very kindly. "Everything has been taken care of. You stay right here, look after Yeshua, and I will go and talk to his mother and his brothers."

The next day Yeshua's brother James came, with his mother, the other brothers and sisters did not believe what Joseph told them. But Mary and James came, and they could not believe their eyes. James behaved rather strangely, he told Joseph, 'What kind of trick was that? This is not my brother!' So Yeshua looked at him and told him, 'Do you remember the cut I had on the upper side of my left shoulder?' I knew he had the cut. So James, said, 'Yes, let me see.' And Yeshua showed him the cut, and James fainted.

Joseph told us that by tomorrow at the latest, we would go to Tyre for a few days, then get a bigger ship and go the Island of Arwad. He had friends there, business associates, dealing with the olives and olive oil business.

I went to Arwad with Yeshua, Nicodemus, and Joseph, while Mary and James went back home to Judea. Joseph told Mary that within a week, or ten days, I don't remember exactly, he would come back to Judea and take her with him to see us somewhere else.

He did not tell her where we would be, just in case if she would be questioned. So Mary, James, and the other brothers and sisters stayed in Judea. We spent three days in Arwad, a beautiful small island. Many Phoenicians were there, of course speaking fluent Aramaic, so we could mingle with the crowd and nobody knew who we were.

Nevertheless, we mostly stayed inside the house Joseph rented for us. We did not feel safe.

Then one afternoon, Yeshua was amusing himself doing one of his old tricks. He was trying to lift himself up in the air, attempting to fly, or float. He did manage to float a little bit. He used to do a lot of tricks like that.

While we were enjoying his attempts, Joseph returned from an errand, and told us that tomorrow we were going to Cyprus, since Arwad was not really safe. Unfortunately, I started to feel extremely tired, and I suspected I was pregnant, but nobody knew; I did not want to worry either Yeshua or Joseph, particularly since we were not married yet.

We were planning on getting married, as we discussed before, but so many things happened to prevent it. However, now, there was not time to lose; we had to be married before the child was born. But first, we had to get to Cyprus. So when we arrived, a few days later, the very first thing we did was to get married, and I felt so much the better for it.

Joseph again rented a house for us, and then he and Nicodemus returned to Judea. Actually, his plan was to bring Mary to us. He was uneasy about her safety there, and so were we. One month later, he indeed brought her to Cyprus, and I was so happy to see her and tell her about our marriage and the coming child.

Joseph stayed a very short time, and then returned again to Judea. He warned us to stay put and wait until we hear from him before we did anything. And so we lived quietly, and Yeshua regained his strength, but he was limping, and could not walk straight without leaning on me as he walked. I realized that his full healing will take some time, but I was very happy in Cyprus, away from the trouble in Judea.

After another month, Joseph returned. I asked him, in confidence, about the disciples. He told me to forget them, never count on any of them. They were cowards. Only Peter showed some regret, visited Joseph once in a while. The others just went on with their lives. I am sorry to say that Yeshua's brothers and sisters also avoided him, pretending they did not know he was alive. Fear would do such things to people.

Eventually I gave birth to my eldest daughter, Sarah. We stayed in Cyprus until Sarah was three years old, and in the meantime Yeshua found jobs here and there. Everybody liked him, but he lived very quietly and did not go out much. Most important, I insisted that he should not make trouble, or start preaching. No more stories or sermons.

The truth is, he did not want to do so anyway. I think he had enough of sermons, disciples, and preaching. Of course, he could no longer do physical work, so there was no carpentry or handyman jobs for him, which made him a little sad because he liked physical labor.

But he put to use his considerable knowledge of languages, and became a scribe.

Once every five or six months, Joseph came to see us. Life was pleasant enough, but Mary never quite recovered from the ordeal of the crucifixion. She was so delicate, and the ordeal broke her health for good. Eventually became very ill, and even though we tried everything to cure her, she died rather suddenly. We buried her in Cyprus, but years later, Joseph took her body back to Judea for her final resting place, next to her husband, who was also called Joseph and who I have never known, to my sorrow, since he died before I met Yeshua. I still miss Mary.

Following Joseph's advice, we decided to go to Gallia Narbonensis, which had a few large Jewish communities. The idea was to settle in Massilia. Joseph went with us, and also an Ethiopian maid we had in Cyprus, a seventeen years old girl that Joseph brought us for help.

She loved us and did not want to part from us. Later, in Massilia, she married a very nice young man, had a family, and we still visit each other. On the way, I asked Joseph why he took so much trouble to help us.

"You went beyond friendship, even beyond family requirement, Joseph," I said to him with gratitude. "Why are you so kind to us?"

Joseph was quiet for a short time, thinking, musing. "I love you like my own family," he said to me, his honest black eyes looking earnestly into mine. "But it is more than that. I have caused Yeshua, his mother, and you, all the suffering that you have undergone. I will never forgive myself, and forever I will have to atone for my sin."

"You? Caused us suffering?" I asked, incredulously.

"Don't you remember?" he said. "I was the one who asked Yeshua the fatal question during the trial. If I had not asked him who he was, they would have set him free."

I had to cry. "No, Joseph, it was not you. Certainly, the question was misjudged. But the Sanhedrin would have tried to kill Yeshua no matter what. Please, please, don't think about it anymore." We never discussed it again, but I don't think I had any luck in changing his mind, the poor, good man that he was.

Joseph, of course, knew many people in Massilia, since he had much business with them. He took us to a small shop owned by a Jewish friend, who had rented a house for us and got it ready. I loved our house from the first day, it felt like home, even the smell of the house was a little like the houses in Judea, for some reason, possibly because we always lived in fishing villages, and Massilia was a fishing town.

For the first time I felt really safe, far from everyone, Caiaphas in particular. You see, until we came here, I was always afraid Caiaphas will hear about us somehow.

He had his spies everywhere. But Gallia Narbonensis was far enough from Judea, and only business people, like Joseph, would have much to do with it.

Yeshua, too, began to get used to the place, and consorted with Jews only, since they were the safest, we thought. He got a job as a scribe. He kept on surprising me. In Cyprus, I saw him writing Greek. I told him I never knew he spoke Greek, and he laughed and said, 'I speak all the languages.'

'I asked him, where did you learn it?' He said, 'I learned it in Qumran, from the Essenes'. So here he developed more skills. He brushed up on his Latin, and started to learn Gaelic, he was so good with languages, so now with Hebrew, Aramaic, Greek, Latin, and Gaelic, business was okay, and life became comfortable.

We had two more children, a boy and a girl. The boy is called Joseph, after Yeshua's father. You can't say he is named also after Joseph of Arimathea, since we are not allowed to call babies after living people. The girl is called Rachel. And you know our eldest is Sarah. They are all good children, I am happy to say."

Mary Magdalene suddenly stopped talking, and I saw she was smiling at someone behind me. I turned my head, and saw a man entering the room.

"Ambar-Anati, this is my husband, Yeshua. Yeshua, this is a new friend, she came all the way to hear your story."

We greeted each other. It was hard to believe that I was looking, meeting, talking with Jesus Christ. This pleasant, ordinary, normal man?

Could it be? He was not at all what I expected. He did not have fair long hair, he did not possess an ascetic, pale face, or an emaciated body. Instead, Yeshua had frizzy dark hair, a little curly on the back. He had dark skin, black eyes, and he was strongly built, stocky, and not very tall, probably five eight or nine; you could say he was a little plump, though certainly not fat, Just comfortable looking.

He wore sandals, a blue gray outfit, and carried two bags. One was a leather book bag, and the other a basket full of eggs and dry fruits. Jesus Christ, shopping at market? And yet, it had to be him, Jesus of Nazareth, Jesus Christ... There could be no one else with a history so similar.

I lost my head and I asked him, "Are you the One?" Fortunately, he did not understand what I meant. He smiled and said, "I am sorry, what did you say?" Stupidly, I asked him again, are you the Messiah? He laughed like a child, and said, "Don't bring up old stories and memories, all this is well behind me..."

Well, at least I did not kneel before him. That would have not gone too well. So I recovered myself and smiled at him, just as if he were a normal new acquaintance.

"Let me make some supper," said Mary Magdalene. "Please stay and eat with us, Ambar-Anati. You will also like to meet the children, I am sure."

"I am sorry, Mary. I really must go, I have to meet someone and go home."

"May I take you where you need to go?" Yeshua asked helpfully.

"No, really, there is no need. I am meeting them by the shore, just a few steps."

"If there is any trouble, though, and they don't arrive on time, come right back, won't you?" said Mary Magdalene.

"I will. And with your permission, I would like to visit again, and meet the children."

"I am counting on it," said Mary Magdalene. We parted cordially, and I left. I was going to the shore, where I could be hidden from sight as I planned to give the code and go back to Nibiru through my Monitor.

As I was turning around the house, slowly, thinking about this wonderful experience, I was passing the window and heard Yeshua's voice. I stood for a moment, listening. I did not want to eavesdrop, it was just that it was hard for me to part from these two wonderful people who meant so much to me, and I just lingered.

"Wait one minute, Mary, before you go to make the supper," Yeshua said. "I have something really interesting to tell you, which I want to do before the children come. I don't want them to hear it. It is downright amazing... there is a Greek man out there in the public plaza, he is doing what I used to do... but he is much smarter than me. He was preaching about God, and of all things, who do you think he was also talking about, to a dozen of people? Me! was talking about me!"

There was a short silence. Then I heard Mary Magdalene say, her voice full of anxiety, even terror. "Yeshua, did you get in trouble? Did you do something stupid?"

"No, no, I did nothing dangerous. I just asked the fishermen, 'Who is this man?' They said he was a Roman, or a Greek, and his name is Paul, or Saul, or something like that... he came from Judea. They said he personally knew a demigod, Jesus, something like that... Well, I had to talk to him, because I know the language, and Jesus is a version of my own name. I was terribly curious."

"You talked to him? He could have gotten you arrested!" cried Mary Magdalene, aghast.

"He is much more likely to be arrested himself, my dear. He is talking blasphemy, while I am just the respectable scribe everyone knows around here... I took him aside and we talked for a short time. I said to him that I think it was I he was talking about, my name is Yeshua.

He looked at me, and his eyes were shifty, mean, and shrewd. I usually like people, as you know, but I did not like this man. He said, 'Maybe you are, maybe you are not, my man. I really don't care either way, because you don't count. I have come with a new religion, I have brought the Messiah. Don't interfere with me – I plan to have a following all over the world.' Now, when he mentioned the word Messiah, I just had to laugh, and so I slapped him on the shoulder and left him alone to continue with his business."

"So did he go back to his preaching?" asked Mary Magdalene. Her voice, I was pleased to tell, was not so strained anymore.

"Yes, he went right back to his preaching," said Yeshua, his voice full of laughter. "I stood at a little distance and listened to him as he talked on and on. Do you know what he said? He was telling everyone that I died on the cross, and then I rose from the dead and came back to life! Even I, in my best days as a magician, could have never invented such nonsense. Rose from the dead! Can you believe it?"

*** *** ***

120. "BEE" Or "EBE"?
⌘ ⌘ ⌘

BEE" or "EBE"?
I. Definition and origin of the terms "BEE" and "EBE"
II. "Presidential Report; Reference: PT/EBE-01"
III. CIA's EBEs file, called "The Bible"

I. Definition and origin of the terms "BEE" and "EBE":

B.E.E. and/or E.B.B. are two identical acronyms for Biological Extraterrestrial Entity, and Extraterrestrial Biological Entity. The first term is commonly and widely used by ufologists, while the second was exclusively used by military physicians for years. In other words, B.E.E., and E.B.B., mean aliens or extraterrestrials. According to one military surgeon, the term BEE was used in the original and master medical files on autopsy(ies) of dead aliens.

The acronym B.E.E. was popularized by William Moore and associates. The Pentagon and the CIA extraterrestrials' classified files did not fall under the classification of "Humanoid", and/or "Humanoid bodies" as claimed by Moore, Jaime Shandera, and associates.

The cabinet files' labels read: "EBE."
The military medical labels read: "BEE".
The staff and/or technicians (Military and medical) referred to the "humanoids" as "AB"; an acronym for aliens bodies.
The word or term frequently used on the floor was AB. (Floor means inside the facility, and/or an area strictly reserved to extraterrestrials' affairs.)
The alleged MJ12 document (Never substantiated) indicates that "the high-profile team directed by MJ12 member and physiologist, Detlev Bronk has originally suggested the term of "Biological Extraterrestrial Entity" (BEE), and recommended that it should be adopted as the standard term of reference for these creatures, until such time, as a more definitive designation

can be agreed upon." But six months later, the term was changed to "Extraterrestrial Biological Entities", or "EBEs".

Brief mention of EBEs was also made in a top secret report on a December 6, 1950 UFO's crash, along the Texas-Mexico borders.
In this report, the five aliens who were found alive on the site of the crash were called "EE", meaning extraterrestrial entities.
In one of the United States Air Force files on EBEs, which was submitted to President Truman, there is one paragraph that lists and describes the crashes and retrievals of UFOs and their occupants.
The occupants of the UFOs were called EBEs, and were described as "3 1/2 to four feet tall, gray-skinned and hairless, with oversized heads, large eyes and no noses."
Furthermore, according to whistleblowers, the document stated the following on a subsequent page, "EBEs from a nearby solar system, have been here on earth for many thousands of years. Through genetic manipulation they influenced the course of human evolution and in a sense created us.

They had also helped shape our religious beliefs."
This particular file was diligently reviewed by President Truman, who simply referred to the aliens' file as the "EBE File". According to a retired US Air Force colonel, who briefly served at The White House, "in all the meetings, the President called them EBE."

According to Bob Lazar, the government documents he reviewed stated that "the aliens or EBEs were from the fourth planet orbiting Zeta 2 Reticuli, the second star of a binary system in the Constellation Reticulum.
A day on Zeta 2 Reticuli Planet 4 is about 90 hours long. The Reticulan EBEs are three to four feet tall and weigh twenty-five to fifty pounds.
The EBEs' bodies vaguely resemble a human toddler's torso if emaciated from hunger.
They have grayish skin and large heads with almond shaped wrap-around eyes. They have very slight nose, mouth, and ear positions and are hairless."

*** *** ***

II. "Presidential Report; Reference: PT/EBE-01":

- a- Another UFOs' crash that occurred at Roswell in 1949.
- b- The names of four military investigators and one civilian scientist who were at the crash site, and who found five EBEs' dead bodies and one alive .
- c- The alive EBE was taken to Los Alamos National Laboratory, located north of Albuquerque.
- d- The "aliens" were called EBEs in the report, not extraterrestrials or aliens, because according to a footnote in the "Presidential Report No. PT/EBE-01" the military did not know the origin of these entities. It was also reported, that President Truman called these aliens "Entities" on more than one occasion.
- e- Part of the anatomical/physical description of the EBES included the following: "They are small and skinny gray humanoids."
- f- The new name given to the EBE who was still alive was EBE-1.
- g- In a third report submitted to the White House, it was mentioned that the EBE (The one who was found alive) died on June 18, 1952.
- h- This report also included a new categorization given to EBES: EBE-2, and EBE-3.
- i- The acronyms EBE-2, and EBE-3 referred to aliens who were living and working with military scientists at a secret military base.

III. CIA's EBEs file, called "The Bible":

It was rumored that at the CIA headquarters in Langley, Virginia, there is a thick book called "The Bible," which is an extensive compilation of all UFOs' crashes, description of EBEs categories, and secret US-aliens project reports.

It seemed that these rumors were created either by Moore, John Lear, or Richard Doty. Others claim, that they came directly from former CIA's agents. But none of these rumors were ever substantiated.

Also in "The Bible" a big file on Bennewitz was found.

The file contained many sketches, charts, photos, and "contact names" of Bennewitz. According to his own account, which he would not reveal until 1989, Moore cooperated with the CIA, NSA, Richard Doty, and AFOSI.

The United States government told Moore that as his part of the deal, he was to spy on Bennewitz, some ufologists, and APRO.

In Moore's words, "to a lesser extent, several other individuals." (Moore, 1989). Moore said that several government agencies were interested in Bennewitz's activities and they wanted to inundate Bennewitz with false "information-disinformation", to confuse him. Moore says he was not one of those providing the disinformation, but he knew some of those of who were, such as Doty. Some insiders have claimed that "The Bible" was very well known to the NSA, which refrred to as "The Lunatics Book."

Note: This is not totally correct, because the "Lunatics Book" was in fact a list of ufologists, the CIA and NASA were watching very closely and very carefully. The "Lunatics Book", was not a book, but simply a list of names.

*** *** ***

121. The Anunnaki Ba'ab "Stargate"

⌘ ⌘ ⌘

Ba'ab "Stargate"
Assyrian/Babylonian/Sumerian/Ana'kh. Term for Anunnaki's stargate.
I. Definition
II. Etymology
Note on the CIA's "Star Gate Program"

I. Definition:

Ba'ab "Bab" is a term for an Anunnaki's stargate; an entrance and exit to multiple worlds, stars, planets, and galaxies. The Anunnaki deploy Ba'ab "split-time-space technology" to travel to any region of the universe in a fraction of a minute.

In contemporary science fiction literature, quantum physics, and ufology, it could be compared to the Stargate technology

This technology allows the Anunnaki to bend time and space and travel to various dimensions, levels and types of the cosmos, such as:

- 1-The multi-dimensional world,
- 2-The parallel world,
- 3-The future world,
- 4-The galatico-plasma world,
- 5-The past dimension,
- 6-The future dimension,
- 7-The multiverse zones,
- 8-The space-memory zone,
- 9-The Akhashic sphere.

This also allows them to send and receive instant knowledge and messages from and to the beginning of the universe, the beginning of time, and the beginning of motion.

This time-space technology is not a monopoly of the Anunnaki.

Many extraterrestrial civilizations are familiar with the concept and its pragmatic application.

II. Etymology:

From Ba'ab "Bab", derived:

❖ **a**-The Sumerian word Babu. (The plural is: Babani and Babati).
❖ **b**-The Arabic word Bab. (The plural is: Abwaab.)

Some ancient Near Eastern and Middle Eastern civilizations incorporated the word Bab in their most important symbols and national entity vocabularies, such as the word of Babylon, which means the gate of God.

Babylon is composed of two words:

❖ **a**-Bab (Gate),
❖ **b**-Ylon or Eli, or El (God).

"Zahi asbut bab-rab-sunnu ana izziqipi uzaqip."
From the Annals of Sardanapalus. Translated verbatim: "700, men about their great gate on crosses, I crucified."
King Esur (Assur, Ashur) said: "usashira gimir babani."
Translated verbatim: "I put around all the gates."
And: "mihrit babi-sin." Translated verbatim: "Before their gates."
From the Annals of Sardanapalus: "Zabi tilai in babatte "Babati" sa er-su ana ziqipi lu uzaqipi."
Translated verbatim: "Men alive at the gates of his city on crosses I impaled."
And from the slabs of Sennacherib: "Mihrit babati."
Translated verbatim: "Before the gates."

Note on the CIA's Star Gate Program:

Many still believe that the CIA's "Star Gate Program" rotates around extraterrestrials' portals, wormholes, and galactic tunnels that open up to allow American astronauts and military space crafts, and vehicles to travel to the edge of the universe.
These assumptions are not accurate.
However, the CIA did work on a project pertaining to space-time intergalactic travel, with the cooperation of the United States Air Force, NASA, NSA, and MIT. This project is explained under the entry "Space-Time Travel".

The CIA's Star Gate was one of a number of "remote viewing programs" conducted under a variety of code names, including SUN STREAK, GRILL FRAME, and CENTER LANE by DIA and INSCOM, and SCANATE by CIA.

***   ***   ***

122. UFOs, Aliens, Aliens' Rapture, Nichola Tesla, and the United States Government
⌘⌘⌘

- I. Definition and introduction
- II. Channeling with extraterrestrials and entities from galactic civilizations is impossible
- III. Any authoritative source(s) on the aliens' rapture and extraterrestrials' communications?
- 1. Introduction
- 2. These sources are
- a- The Anunnaki-Ulema
- b- The Book of Ramadosh
- c- Statements by Anunnaki-Human-Hybrids
- d- Statements by a couple of highly respected astronomers
- e- Opinions of bona fide scientists
- f- Opinions and claims of eminent scientists and professors at Moscow Academy of Medicine (Moscow Sechenov Academy of Medicine)
- g- A World War Two famous cryptologist who worked on the German "Enigma Code Machine"
- h- The personal files of Nichola Tesla
- Nicola Tesla's files on extraterrestrials
- Foxworth, then, assistant director of the New York FBI bureau was called in
- J. Edgar Hoover, Dr. Trump and Tesla's files
- The "scary report" of Dr. Trump
- Clyde A. Tolson hides Dr. Trump's report
- Hoover falsified/switched Dr. Trump's report
- Dr. Trump's conclusions
- Tesla, Brigadier General L. C. Craigie, and Patterson Air Force Base, and the "Project Nick."
- Lt. Gen. Laurence C. Craigie's Promise to Truman

- More scientists, including Fermi were called in
- "Project Nick" is now "Project Lexicon"!
- Extraterrestrials' warning
- What's left from Tesla's files are now in the hands of the Yugoslavian army
- Soviet Premier Nikita Khrushchev's announcement to the Supreme Soviet (Presidium).
- Tesla was concerned about the "Safety of Earth"!
- Sava Kosanovic taken under custody for two full weeks
- Further investigations of Tesla's scenario of the alien rapture
- President Eisenhower allocated a huge budget
- "Project Seesaw"
- The Russians built their first Tesla's beam weapon facility
- The CIA learned about the Russian facility from alien space-beacons orbiting Earth
- The "Extraterrestrial Officers"
- New and mysterious kind of satellite totally unknown to scientists.
- The Pentagon database, and the US Navy secret file on "Non-Terrestrial Officers"
- Actual sounds from these space machines
- Hacker uncovered US off-planet space navy
- Russian Telsa's beam weapon system and the United States Strategic Defense Initiative (SDI)
- President Reagan on the "Aliens' Threat"
- The FBI secret files on Tesla

*** *** ***

122. UFOs, Aliens, Aliens' Rapture, Nichola Tesla, and the United States Government

I. Definition and introduction:

What is the aliens' rapture?

It is part confrontation/collision, and part separation.

Although extraterrestrials are highly advanced, their high level of technology and baffling scientific knowledge are not categorically a code for morality and ethics.

Science, whether on planet Earth or beyond, can be used for ill purposes. Such possibility exists all over the universe.

Thus, frictions, confrontations, oppositions, separation, and wars on the galactic landscape of extraterrestrials happen on a regular basis.

In fact, the very first time, humans have heard of visitors from outer space, was through depictions of aliens' destruction of cities on Earth, and wars between God and the Fallen Angels, and between the Nephilim, the Biblical Giants, the Watchers, the Guardians, the Legions of Darkness, and the Shining Ones, in the Bible, in the holy scriptures, in the banned books from the Bible, in the Gnostic scrolls, and ancient Mesopotamian clay tablets. And all them were of an extraterrestrial stock! Aliens' destruction of cities, burning people alive, and turning humans into pillars of salt were neither promising, nor reassuring for humankind. So yes, extraterrestrials can fight each other, as we do often here on Earth.

II. Channeling with extraterrestrials and entities from galactic civilizations is impossible:

Of course, we don't know much about the social structures and social values of highly advanced beings on other planets.

Basically, what ufology's literature provides on the subject comes from channelers, experiencers, and contactees who allegedly met extraterrestrials, and received their "spiritual" messages through telepathy, as they often claim.

For zillions of reasons. I do not take seriously these messages and channelings affairs.

Mainly because, contacts with extraterrestrials via channeling is impossible. And if such channeling does in fact exist, the human brain' frequencies and mental communications do not cross the frontiers of the solar system.

Unless, those channelers are using some sort of mind-boggling interstellar beams, and/or intergalactic mental/radio waves faster than light.

Even light bends on itself, gets distorted, and dissipates in the no-time-no-space zones. Grosso modo, channeling (from Earth) with extraterrestrials and entities from galactic civilizations is impossible."

III. Any authoritative source(s) on the aliens' rapture and extraterrestrials' communications?

1. Introduction:

If this is the case, then we have no way at all of knowing what is going on in other planets, and particularly about duels, feuds, and rapture among extraterrestrial civilizations.

That is correct.

However, there are obscure, authoritative and very few sources that tell us something about the aliens' rapture, its origin, past or current development, and how it could affect the present and future of humanity.

These sources also reveal baffling governmental "notes", and top secret memoranda on extraterrestrials' communications between two (or more) alien spaceships.

2. Some of these sources are:
- **a**-The Anunnaki-Ulema.
- **b**-The Book of Ramadosh.
- **c**-Statements by Anunnaki-Human-Hybrids.
- **d**-Statements by two highly respected astronomers, currently teaching at leading American universities, and working at some of the world's most advanced observatories.

Their names will not be released, even though, they gave their statements on the History Channel and the Science Channel in the United States, in 2008.

- **e**-Opinions of bona fide scientists (Astrophysicists, cosmologists and astronomers) who worked at SETI, for a very long time.

Their names will not be released, even though, they gave their statements in open discussions, and on the History Channel and the Science Channel in the United States, in 2008.

- **f**-Opinions and claims of eminent scientists and professors at Moscow Academy of Medicine (Moscow Sechenov Academy of Medicine), given in 1987, and in 1988.

Their names will not be released, as long as, they are still teaching at the academy.

- **g**-A World War Two famous cryptologist who worked on the German "Enigma Code Machine", cipher machines, and teleprinters, previously located in Room 40 of the British Admiralty Building, an undisclosed unit of Britain's military intelligence. In 1957, he explained how the British in 1956 and in 1959, tried to decipher some extraterrestrial messages.

For national security reasons, his name shall not be released.

- **h**-The personal files of Nichola Tesla, seized by the United States government, right after his death in 1943. Those files were nicknamed by researchers "The Missing Files".

Nicola Tesla's files on extraterrestrials:

One of the most sensitive and frightening topics involving Nikola Tesla, was his "Death Beam" invention. The FBI and the United States military intelligence units/branches became extremely interested in his invention, and its practical use as a weapon.

The White House and the Pentagon feared that Tesla' beam blueprints might fall into the hands of the Soviets. But his files included more than the blueprints of the "Death Beam". Special agent "A", and Special Agent "C" stumbled on a Tesla's file called "Exotericon Gama".

On the back of the file, the word "Government" was written in blue ink. At that time, nobody took the Exotericon very seriously. The top brass at the Pentagon thought Tesla is "Nuts", despite the fact, that they were fully aware that some of Tesla's undisclosed inventions could change the face and history of modern weaponry.

Written in code, Tesla described in the Exotericon (Later, to be called Tesla Government File):

- **1.** How extraterrestrials communicate;
- **2.** The sequences of dots, "1", and "O" in the extraterrestrials' messages;
- **3.** A warning sent by an extraterrestrial civilization to another galactic colony.

At the end of World War Two, the United States government (Possibly, President Truman's Secretary of State, General George C. Marshall was in charge of the secret correspondence with the British) asked the help of the decipherers at the British Admiralty to decipher Tesla's code, related to aliens' recorded messages by the elusive genius.

In a documentary on Tesla's legacy, PBS stated, word for word, and unedited: "The morning after the inventor's death, his nephew Sava Kosanovic hurried to his uncle's room at the Hotel New Yorker, in New York. He was an up-and-coming Yugoslav official with suspected connections to the communist party in his country. By the time he arrived, Tesla's body had already been removed, and Kosanovic suspected that someone had already gone through his uncle's effects.

Technical papers were missing as well as a black notebook he knew Tesla kept—a notebook with several hundred pages, some of which were marked "Government."

Note: The notebook marked "Government" is the very same file originally called by Tesla "Exotericon Gama".

Foxworth, then, assistant director of the New York FBI bureau was called in:

P. E. Foxworth, then, assistant director of the New York FBI bureau, was called in by Hoover to investigate, to gather all Tesla's scientific documents (Available and missing) and to come up with a list of Tesla's friends, especially those who were know to have a tie with the communists.

Back then, Foxworth stated that the United States government was "vitally interested in preserving Tesla's papers...for posterity...and further scientific studies..."

Now we know, the United States government's interest in preserving Tesla's papers was neither of an academic nature, nor the preservation of the work of a genius.

My sources tell me that "Just two days after Nicola Tesla's death in New York, four officials from the Office of Alien Property (OAP) went to his room at the New Yorker Hotel and confiscated everything Tesla had in his room.

Absolutely everything, including his "robe-de-chambre" and shoes."

The FBI and OAP used the word "possessions", instead.

Author's note: The word "Possessions" means all his possessions, including all personal effects, scientific drawings, formulae, and particularly the "Beam Death" file (N. Tesla Death Ray as later called by the FBI), and the "Government" file.

J. Edgar Hoover, Dr. Trump and Tesla's files:

The "scary report" of Dr. Trump.

J. Edgar Hoover, then, director of the FBI called Dr. John G. Trump, a scientist (An electrical engineer who worked at the National Defense Research Committee, established by the Office of Scientific Research and Development "OSRD"), and asked him to study and assess the two files in question.

At that time, the OAP has already sent to Hoover, a duplicate of those files. My sources tell me, that Dr. Trump provided Hoover with a "scary report". This is exactly how Trump's report was viewed and called by Hoover himself, and the director of the FBI bureau in New York.

Clyde A. Tolson hides Dr. Trump's report.

Hoover, as usual, a smooth talker and a master of disinformation and threats, kept the original report of Trump in the secret vaults of the FBI. Fearing that something might happen to the secret vaults, Hooper asked his close friend Clyde A. Tolson to hide Trump's file somewhere nobody could find.

Later, rumors surfaced that Tolson kept Trump's report in the attic of an aunt of his who lived on Wisconsin Avenue, Northwest, Washington, DC.

Hoover falsified/switched Dr. Trump's report.
Pressed by the Pentagon, OSRD, and OAP to release the report of
Dr. Trump, Hoover at once, summoned Trump and asked him to
come up with a new version of his report, and to delete any
references made to extraterrestrials, and to the "Ray of Death".
A deja-vu camouflage that echoes the infamous fake photos of
the Roswell's "Weather balloon" wreckages, manufactured and
displayed by the military.

Dr. Trump's conclusions:
Here are Dr. Trump's conclusions: "His thoughts and efforts
(Meaning Nicola Tesla) during at least the past 15 years were
primarily of a speculative, philosophical, and somewhat
promotional character often concerned with the production and
wireless transmission of power; but did not include new, sound,
workable principles or methods for realizing such results." And
that was the end of the story and Tesla's files saga.
But was it? No!

**Tesla, Brigadier General L. C. Craigie, and Patterson Air
Force Base, and the "Project Nick."**
My sources tell me that around November 17 or 19, 1947, the
Pentagon called Brigadier General L. C. Craigie to inform him,
that they will be sending him Tesla's "Government File"
(Exotericon), and "Tesla Beam File" for further study.
The big boys at the Pentagon became convinced that the
extraterrestrials code deciphered by Tesla could help them solve
some of the mysteries of the alien technology and the "Beam
Stick" they found on the site of a UFO crashed near Roswell.
On November 21th or 22nd, 1947, the files were dispatched to
Patterson Air Force Base in Dayton, Ohio, where Brigadier
General L. C. Craigie started a secret operation code-named
"Project Nick". Nick refers to Nicola Tesla.
Dr. Teller and von Braun were called in to participate in the
operation, and lend their scientific expertise.
Craigie began a series of experiments with no apparent success.

Note on Brigadier General L. C. Craigie:
Lt. Gen. Laurence C. Craigie's Promise to Truman:
Reporter Billy Cox has reported that there may have been yet
another General sent to Roswell following the July 1947 crash.

That General was Major General L. C. Craigie. Although he never disclosed what he discovered, according to his personal pilot, he promised then President Harry Truman that he wouldn't talk about what occurred in Roswell.

The official Truman calendar shows the only times Truman met Craigie was at the presentation of the Collier Trophy to Lewis A. Rodert, Chief of the Flight Research Section at the Cleveland Laboratory of the National Advisory Committee for Aeronautics on December 17, 1947.

Craigie was also famous in that he sat on the JRDB which was headed up at the time by Dr. Vannevar Bush. Dr. Bush was described in a Top Secret Canadian memo as the head of a "small group looking into UFOs.

The Research and development Board also had many other key figures on its board who have been tied into the original effort to coverup what was known about flying saucers in the late 1940s. This follows the same pattern as Edwin Easley, who as Provost Marshall for the Army Air Force was in charge of security and clean-up at the Roswell site.

His daughter Nancy Easley Johnson, stated on the July 1, 2003 Larry King Live Show that her father had promised not to talk about what had happened in Roswell after making a promise to President Truman. On his deathbed, Easley finally told his two daughters that he had seen "creatures" at the Roswell crash site. Reported by Billy Cox, in the Herald Tribune.

More scientists, including Fermi were called in.

One week later, Dr. Fermi was called in. On January 17, 1948, an Italian scientist who apparently has worked on some projects started either by Guglielmo Marconi and/or his associates was flown from Rome to Dayton, Ohio.

Around June 1948, the military discontinued "Project Nick". And that was the end of the Tesla's files on extraterrestrials and ray of death. Really?

Not really!!

"Project Nick" is now "Project Lexicon"!

Few things we know for sure:

    a- All the files of Tesla vanished from the face of the earth.

b- The military instructed Teller, von Braun, Fermi, and the "Marconi Guy" (As he was called by peers) to keep their "mouth shut."

c- The military continued "Project Nick" under a new code-name: "Project Lexicon"!!!!! Wow! Very a propos, considering the military interest in deciphering the extraterrestrials' language, coded messages between two alien spaceships, and above all, the warning the military received from aliens they have apparently met on two occasions on a military base.

Extraterrestrials' warning.

From the extraterrestrial warning, some military scientists concluded that the extraterrestrials are fighting each other, and some alien species are already here on Earth, preparing for a major confrontation.

What kind of confrontation?

Nobody knew at the time. However, in a third meeting with the aliens, the military learned that a "Gray Alien" species living underwater, and in an adjacent dimension, will confront the Americans, unless an agreement is reached between the aliens and the United States government.

The aliens did in fact refer to an "Alien Rapture" that has occurred under the nose of an "Extraterrestrial Council", commonly known in ufology's literature as the "Federation".

What's left from Tesla's files are now in the hands of the Yugoslavian army.

My sources tell me that in 1952, and upon the continuous requests and demands of Yugoslavia's Tito to release Tesla's papers, a bulk of Tesla's files and some personal effects were released to Sava Kosanovic. A colonel from the Yugoslavian army met with Kosanovic to take possession of Tesla's papers.

Reluctantly, Sava Kosanovic released Tesla's "remaining" papers to the Yugoslavian colonel.

For years, no scientist or journalist from the West was allowed to see Tesla's papers. And nobody knew what kind of secrets and/or military weapons blueprints Tesla's papers contained.

But everything changed in 1950, when a scientific committee in Belgrade announced, that Tesla's papers will be finally displayed in public, and the scientific community will gain direct access to

Tesla drawings and blueprints of his scientific discoveries, and inventions.

Soviet Premier Nikita Khrushchev's announcement to the Supreme Soviet (Presidium).

In 1960, Soviet Premier Nikita Khrushchev told the Supreme Soviet that "A new and fantastic weapon was in the hatching stage." No doubt, he was referring to Tesla's inventions. Some double agents claimed that Khrushchev was bluffing as usual.

Tesla was concerned about the "Safety of Earth"!

In 1953, Sava Kosanovic told a close friend that Tesla on two occasions showed a serious concern about the "safety of earth". According to Kosanovic, Tesla told him that the world is no longer a safe place, and that he was working on a weapon system to counter attack any alien hostile invasion of earth.

Tesla also told him, that he has learned from an intercepted extraterrestrial communication between two alien races, that the aliens are fighting among each other, and one race is intending on colonizing earth. The rapture in the Council of Extraterrestrial Colonies has divided the aliens, and a hostile alien race which is no longer a member of the Council will invade earth and enslave all of us.

Sava Kosanovic taken under custody for two full weeks.

My sources tell me, that the military and intelligence agencies in the United States learned about what Kosanovic has said to his friend, and took him under custody for two full weeks. And what happened to Kosanovic after his detention?

Nobody knows, because Kosanovic has vanished from the face of the earth!

Further investigations of Tesla's scenario of the alien rapture.

The Tesla's "Aliens' Rapture" scenario was neither disregarded nor ignored by the military and intelligence community. Further investigations of the "Rapture", and Tesla's "Death Ray" (Beam Weapon) continued until 1958.

President Eisenhower allocated a huge budget.

In fact, before the end of 1958, President Eisenhower, secretly, and without telling the Congress, allocated a huge budget for a

Top Secret project, requiring the creation of a military base for the purposes of:

- **a**-Studying and developing a charged-particle beam weapon;
- **b**-Developing a military defense strategy to face any possible alien invasion caused by the "Aliens' Rapture";
- **c**-Exploring the possibility of entering into a "final and friendly agreement with aliens" to prevent any clashes with hostile aliens.

Note: The 1958 original blueprints and specs of Eisenhower's proposed military base were used in their entirety, when the military built AUTEC base in the Bahamas. Of course, the latest weaponry systems, and the world's most advanced technology were added to the repertoire of AUTEC.

"Project Seesaw".
In 1958, the Defense Advanced Research Projects Agency (DARPA) began to work on a top secret project code-named "Seesaw".
Phase one of the research started at the secret military base. Phase two and phase three continued at the Lawrence Livermore Laboratory.
During all that time, military scientists who worked on these "Black Projects" were fully aware of Tesla's papers, including the threatening "Aliens' Rapture."

The Russians built their first Tesla's beam weapon facility.
In October 1958, the Soviets who had access to some of Tesla's "Aliens Rapture" (Government File), smuggled to them by a double agent in Washington, and another major spy in Istanbul, Turkey, began to work on similar projects. Fortunately, nothing came out of it.
But in November 1970, the Russians (A joint team of scientists from Russia, Armenia and the Ukraine) built their first Tesla's beam weapon facility in the Sino-Soviet border in Southern Russia.

The CIA learned about the Russian facility from alien space-beacons orbiting Earth.

In February 1971, the Americans through their double agents in Kiev and Moscow learned about the Russian facility. This was the version of the CIA. (Yet, never disclosed or publicly admitted!) However, according to my sources, the CIA gathered intelligence on the Russian facility and the development of Tesla's beam weapon program, from "friendly" alien space-beacons orbiting Earth.

Some of these space-beacons were piloted by "US Extraterrestrial Officers".

To many of you, this story is absurd and far-fetched. But years later, the "absurd" story of the "American Extraterrestrial Officers" was fully documented and proven real.

Here is the full story.

The "Extraterrestrial Officers":

It seems that some of the stars above us are not stars at all.

Whistleblowers and so-called Pentagon's insiders claim that many of those stars are extraterrestrial ships manned by "Extraterrestrial Officers".

A young man by the name of John Lenard Walson has discovered a new way to extend the capabilities of small telescopes and has been able to achieve optical resolutions - at almost the diffraction limit - not commonly achievable. With this new-found ability, he has proceeded to videotape, night and day, many strange and heretofore unseen objects in earth orbit.

The resulting astrophotographic video footage has revealed a raft of machines, hardware, satellites, spacecraft and possibly spaceships which otherwise appear as stars, if they appear at all. There are, indeed, hundreds of satellites in Earth's orbit.

The Union of Concerned Scientists stated that there are more than 800 active satellites currently in orbit. Amazingly, they represent four percent of the total number of objects currently cataloged by the U.S. space surveillance network; the rest includes abandoned satellites, spent rocket boosters, and other debris.

The United States owns more than 400 active satellites, just over 50 percent of all satellites.

Russia and China have the second and third highest number of space assets, owning 89 and 35 satellites, respectively. Civilian satellites, which perform tasks for the commercial, scientific, and government sectors, make up the majority of U.S. satellites.

Russia's space assets are split nearly evenly between military and civil missions, though there are not separate military and civilian space programs. Only a very small percentage of other countries' satellites are military in nature. Approximately two-thirds of all active satellites are used for communications.

Satellites for navigation, military surveillance, Earth observation and remote sensing, astrophysics and space physics, and Earth science and meteorology missions each comprise about five to seven percent of total satellites. This is the official version.

New and mysterious kind of satellite totally unknown to scientists.

However, the videos of Walson have revealed a new and mysterious kind of satellite totally unknown to scientists. One writer asked: "Are We Supposed To Be Seeing These Things?" Walson received the following comment about one of the videos. Does it answer all the questions?

No, but perhaps some of them.

"Hello again. And, again, my congratulations on your superb astrophotography. MIT Lincoln Laboratory is the group which has built some of the things you are seeing.

Much of what they do is what used to be the Star Wars project, which no doubt involves some of your objects..." this was a message Walson received from scientists at MIT.

Maybe some of Walson's images are of sensitive, secret US military Star Wars machines.

Maybe even secret weapons platforms in space, which the US military has been rumored to have for at least 20 years.

Many scientists are not convinced.

The Pentagon database, and the US Navy secret file on "Non-Terrestrial Officers".

A few years ago, a young UK hacker, by the name of Gary McKinnon, got into the Pentagon database, and found a secret US Navy file titled "Non-Terrestrial Officers".

Someone isn't happy with John Lenard Walson: Judging by the video footage taken by Walson at his home, it is clear he has ventured into territory that is, shall we say, 'disconcerting' to some government or military agencies who don't like him videotaping these large, mysterious orbiting machines.

The result of Walson's 'impertinence' was the almost immediate inauguration of routine visits from numerous unmarked large black helicopters which began regular day and night visits to his home.

Walson was able to capture and display still frames from video he recorded of his tormentors. One of them is the photo of a huge, twin-rotor Chinooks passing right over the tree in his backyard.

Actual sounds from these space machines:

In addition to discovering and refining his optical telescope videotaping technique, John Walson has also discovered how to actually hear and record the sounds in real time coming from the particular craft he is videotaping.

By carefully aligning a satellite dish receiver with his telescope, he has been able to record some very unusual and intriguing sound from the different spacecrafts.

Some ufologists claimed that these mysterious objects are extraterrestrial spaceships, parked or stationed in orbit above the Earth, and "sort of crafts co-manufactured by aliens, NASA and the United Air Forces as anti-satellite weapons."

Most recently, China has been discussing its anti-sat programs and even threatened to destroy or disable all GPS satellites which overfly Chinese territory.

It is anyone's guess how many billions the US military and government have poured into black operations NASA-facilitated programs of 'secret' spying and surveillance programs, and military space weapons systems over the decades. John Lenard Walson's video images and sounds may well show some of these advanced machines.

It is also within the realm of possibility, some of these items might be products of non-human intelligence. Of course, the people who would know aren't talking. Reported in Dark Government; Black Projects; H. Walson; Blip TV; Above Space and Time.

Hacker uncovered US off-planet space navy:

Under that headline, John Rense said (As is and unedited): "This is possibly a profoundly important development.

If Mr. McKinnon's data and assumption are correct, it validates what I and others have been postulating for many years that the US Navy/Military may well be operating off-planet via back-engineered ET technology (or WWII German?) for a long

299

time...long enough to have a 'fleet' of space craft and officers to either man them or otherwise control them.

For those who remember the Clementine mission, you will recall it was a US Navy project which micro-mapped the entire Moon. If McKinnon stumbled onto a secret file of "Non-Terrestrial Officers", it would, indeed, suggest the US Military has been quietly, efficiently, secretly running off-planet operations for a long time."

Gary McKinnon's statement:

Gary McKinnon, the English hacker facing 70 years in U.S. prison for searching Pentagon sites for UFO evidence, says the weirdest thing he found was a list of "Non-Terrestrial Officers" and fleet transfers between ships that don't exist in the U.S. Navy. He said: "I found a list of officers' names, under the heading Non-Terrestrial Officers." He added: "Non-Terrestrial Officers? Yeah, I looked it up, and it's nowhere. It doesn't mean little green men.

What I think it means is not earth-based. I found a list of fleet-to-fleet transfers, and a list of ship names. I looked them up. They weren't U.S. navy ships. What I saw made me believe they have some kind of spaceship, off-planet."

Russian Telsa's beam weapon system and the United States Strategic Defense Initiative (SDI).

Back to the Russian Telsa's beam weapon system and "Aliens' Rapture":

The American instant response to the Russian program was the Strategic Defense Initiative (SDI) proudly announced to the world by President Ronald Reagan in 1983.

He stated: "Teams of government scientists were urged to turn their great talents now to the cause of mankind and world peace, to give us the means of rendering these nuclear weapons impotent and obsolete."

⌘⌘⌘

President Reagan on the "Aliens' Threat".

And herewith for the record, United States concerns about an alien invasion caused by an alien rapture.

The Christian Science Monitor stated: "Former President of the Soviet Union Mikhail Gorbachev spent an hour with Charlie Rose and Reagan Secretary of State George Shultz recently.

The three discussed many topics including a conversation between Reagan and Gorbachev about uniting against visitors from outer space."

 The Monitor's Jimmy Orr wrote: "Reagan and Gorbachev agreed to fight UFOs. This is a perfect one for Friday night. Charlie Rose had the former President of the Soviet Union Mikhail Gorbachev and President Reagan's Secretary of State George Shultz on his program this past Tuesday night.

This year marks the 20th anniversary of the fall of the Berlin Wall. Gorbachev got a laugh when he said he wasn't all that impressed with Reagan's historic challenge to "tear down this wall".

But went on to say that he believed Reagan "was a great president." Shultz was talking about the Lake Geneva summit and mentioned the two leaders ducked out of a meeting to take a walk to a nearby cabin."

I wasn't there...," Shultz said before Gorbachev cut him off." From the fireside house, President Reagan suddenly said to me, 'What would you do if the United States were suddenly attacked by someone from outer space? Would you help us?'

"I said, 'No doubt about it.'" He said, "We too."

And before the United Nations General Assembly, September 21st, 1987, President Reagan said: "In our obsession with antagonisms of the moment, we often forget how much unites all the members of humanity.

Perhaps we need some outside, universal threat to make us recognize this common bond. I occasionally think how quickly our differences worldwide would vanish if we were facing an alien threat from outside this world. And yet, I ask you, is not an alien threat already among us?"

*** *** ***

The FBI secret files on Tesla.

Document 1: Full page

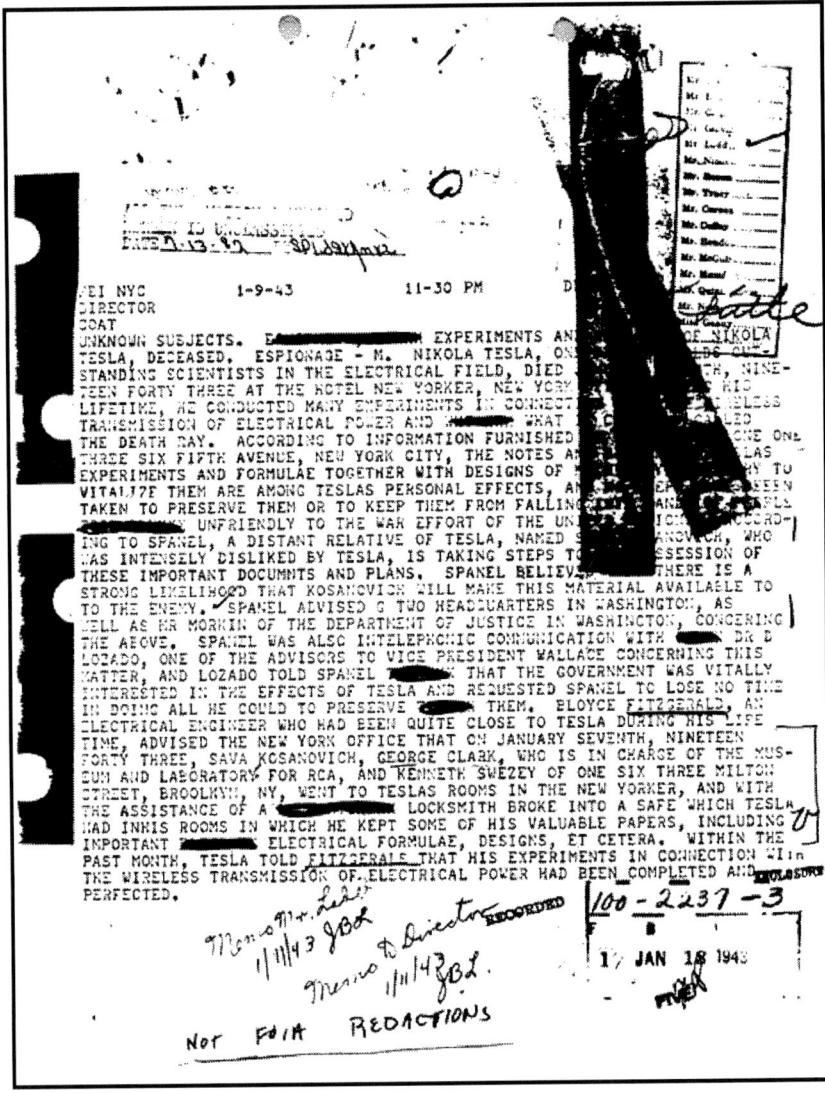

FBI NYC   1-9-43   11-30 PM

DIRECTOR

COAT

UNKNOWN SUBJECTS. [REDACTED] EXPERIMENTS AN[D] [REDACTED] OF NIKOLA TESLA, DECEASED. ESPIONAGE - M. NIKOLA TESLA, ON[E OF THE OUT]STANDING SCIENTISTS IN THE ELECTRICAL FIELD, DIED [REDACTED] NINE-TEEN FORTY THREE AT THE HOTEL NEW YORKER, NEW YOR[K] [REDACTED] LIFETIME, HE CONDUCTED MANY EXPERIMENTS IN CONNECT[ION] [REDACTED] TRANSMISSION OF ELECTRICAL POWER AND [REDACTED] WHAT [REDACTED] THE DEATH RAY. ACCORDING TO INFORMATION FURNISHED [REDACTED] ONE ONE THREE SIX FIFTH AVENUE, NEW YORK CITY, THE NOTES A[REDACTED] EXPERIMENTS AND FORMULAE TOGETHER WITH DESIGNS OF [REDACTED] VITALIZE THEM ARE AMONG TESLAS PERSONAL EFFECTS, A[REDACTED] TAKEN TO PRESERVE THEM OR TO KEEP THEM FROM FALLING [REDACTED] [REDACTED] UNFRIENDLY TO THE WAR EFFORT OF THE UN[REDACTED]CORD-ING TO SPANEL, A DISTANT RELATIVE OF TESLA, NAMED S[REDACTED]NOVICH, WHO WAS INTENSELY DISLIKED BY TESLA, IS TAKING STEPS TO [REDACTED]SSESSION OF THESE IMPORTANT DOCUMNTS AND PLANS. SPANEL BELIEVE[REDACTED] THERE IS A STRONG LIKELIHOOD THAT KOSANOVICH WILL MAKE THIS MATERIAL AVAILABLE TO TO THE ENEMY. SPANEL ADVISED G TWO HEADQUARTERS IN WASHINGTON, AS WELL AS MR MORKIN OF THE DEPARTMENT OF JUSTICE IN WASHINGTON, CONCERNING THE ABOVE. SPANEL WAS ALSO INTELEPHONIC COMMUNICATION WITH [REDACTED] DR D LOZADO, ONE OF THE ADVISORS TO VICE PRESIDENT WALLACE CONCERNING THIS MATTER, AND LOZADO TOLD SPANEL [REDACTED] THAT THE GOVERNMENT WAS VITALLY INTERESTED IN THE EFFECTS OF TESLA AND REQUESTED SPANEL TO LOSE NO TIME IN DOING ALL HE COULD TO PRESERVE [REDACTED] THEM. BLOYCE FITZGERALD, AN ELECTRICAL ENGINEER WHO HAD BEEN QUITE CLOSE TO TESLA DURING HIS LIFE TIME, ADVISED THE NEW YORK OFFICE THAT ON JANUARY SEVENTH, NINETEEN FORTY THREE, SAVA KOSANOVICH, GEORGE CLARK, WHO IS IN CHARGE OF THE MUSEUM AND LABORATORY FOR RCA, AND KENNETH SWEZEY OF ONE SIX THREE MILTON STREET, BROOKLYN, NY, WENT TO TESLAS ROOMS IN THE NEW YORKER, AND WITH THE ASSISTANCE OF A [REDACTED] LOCKSMITH BROKE INTO A SAFE WHICH TESLA HAD IN HIS ROOMS IN WHICH HE KEPT SOME OF HIS VALUABLE PAPERS, INCLUDING IMPORTANT [REDACTED] ELECTRICAL FORMULAE, DESIGNS, ET CETERA. WITHIN THE PAST MONTH, TESLA TOLD FITZGERALD THAT HIS EXPERIMENTS IN CONNECTION WITH THE WIRELESS TRANSMISSION OF ELECTRICAL POWER HAD BEEN COMPLETED AND ENCLOSURE PERFECTED.

100 - 2237 - 3

17 JAN 18 1943

NOT FOIA REDACTIONS

Enlarged: Document 1

First half of the page

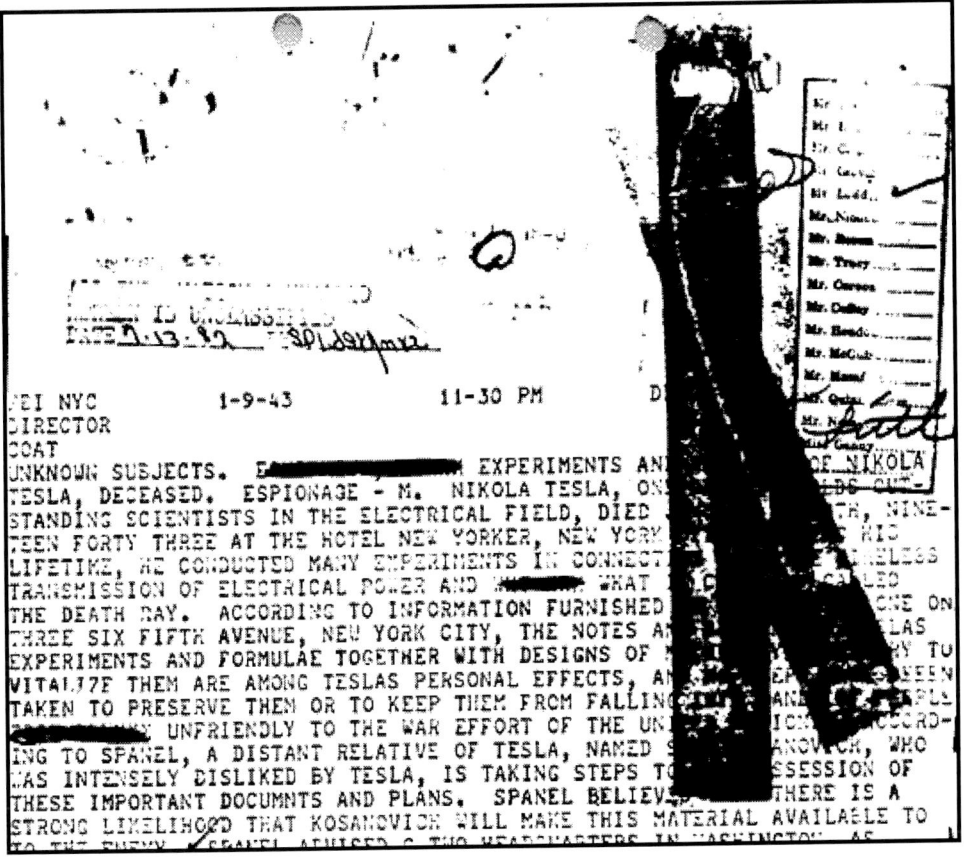

EI NYC   1-9-43   11-30 PM
DIRECTOR
COAT
UNKNOWN SUBJECTS. ▬▬▬▬▬▬▬▬ EXPERIMENTS AND
TESLA, DECEASED. ESPIONAGE - M. NIKOLA TESLA, ONE
STANDING SCIENTISTS IN THE ELECTRICAL FIELD, DIED
TEEN FORTY THREE AT THE HOTEL NEW YORKER, NEW YORK
LIFETIME, HE CONDUCTED MANY EXPERIMENTS IN CONNECT
TRANSMISSION OF ELECTRICAL POWER AND ▬▬▬▬ WHAT
THE DEATH RAY. ACCORDING TO INFORMATION FURNISHED
THREE SIX FIFTH AVENUE, NEW YORK CITY, THE NOTES A
EXPERIMENTS AND FORMULAE TOGETHER WITH DESIGNS OF
VITALIZE THEM ARE AMONG TESLAS PERSONAL EFFECTS, AN
TAKEN TO PRESERVE THEM OR TO KEEP THEM FROM FALLING
▬▬▬▬▬ UNFRIENDLY TO THE WAR EFFORT OF THE UN
ING TO SPANEL, A DISTANT RELATIVE OF TESLA, NAMED
WAS INTENSELY DISLIKED BY TESLA, IS TAKING STEPS TO
THESE IMPORTANT DOCUMNTS AND PLANS. SPANEL BELIEV
STRONG LIKELIHOOD THAT KOSANOVICH WILL MAKE THIS MATERIAL AVAILABLE TO

*** *** ***

303

STRONG LIKELIHOOD THAT KOSANOVICH WILL MAKE THIS MATERIAL AVAILABLE TO
TO THE ENEMY. SPANEL ADVISED G TWO HEADQUARTERS IN WASHINGTON, AS
WELL AS MR MORKIN OF THE DEPARTMENT OF JUSTICE IN WASHINGTON, CONCERNING
THE ABOVE. SPANEL WAS ALSO IN TELEPHONIC COMMUNICATION WITH ▓▓▓ DR D
LOZADO, ONE OF THE ADVISORS TO VICE PRESIDENT WALLACE CONCERNING THIS
MATTER, AND LOZADO TOLD SPANEL ▓▓▓▓▓ THAT THE GOVERNMENT WAS VITALLY
INTERESTED IN THE EFFECTS OF TESLA AND REQUESTED SPANEL TO LOSE NO TIME
IN DOING ALL HE COULD TO PRESERVE ▓▓▓ THEM. BLOYCE FITZGERALD, AN
ELECTRICAL ENGINEER WHO HAD BEEN QUITE CLOSE TO TESLA DURING HIS LIFE
TIME, ADVISED THE NEW YORK OFFICE THAT ON JANUARY SEVENTH, NINETEEN
FORTY THREE, SAVA KOSANOVICH, GEORGE CLARK, WHO IS IN CHARGE OF THE MUS-
EUM AND LABORATORY FOR RCA, AND KENNETH SWEZEY OF ONE SIX THREE MILTON
STREET, BROOKLYN, NY, WENT TO TESLAS ROOMS IN THE NEW YORKER, AND WITH
THE ASSISTANCE OF A ▓▓▓▓▓▓▓ LOCKSMITH BROKE INTO A SAFE WHICH TESLA
HAD IN HIS ROOMS IN WHICH HE KEPT SOME OF HIS VALUABLE PAPERS, INCLUDING
IMPORTANT ▓▓▓▓▓ ELECTRICAL FORMULAE, DESIGNS, ET CETERA. WITHIN THE
PAST MONTH, TESLA TOLD FITZGERALD THAT HIS EXPERIMENTS IN CONNECTION WITH
THE WIRELESS TRANSMISSION OF ELECTRICAL POWER HAD BEEN COMPLETED AND ▓▓▓▓
PERFECTED.

Memo Mr. Ladd 1/11/43 JBL

Memo to Director RECORDED 100 - 2237 - 3

Memo 1/11/43 JBL

17 JAN 18 1943

Not FOIA REDACTIONS

*** *** ***

Document 2: Full page

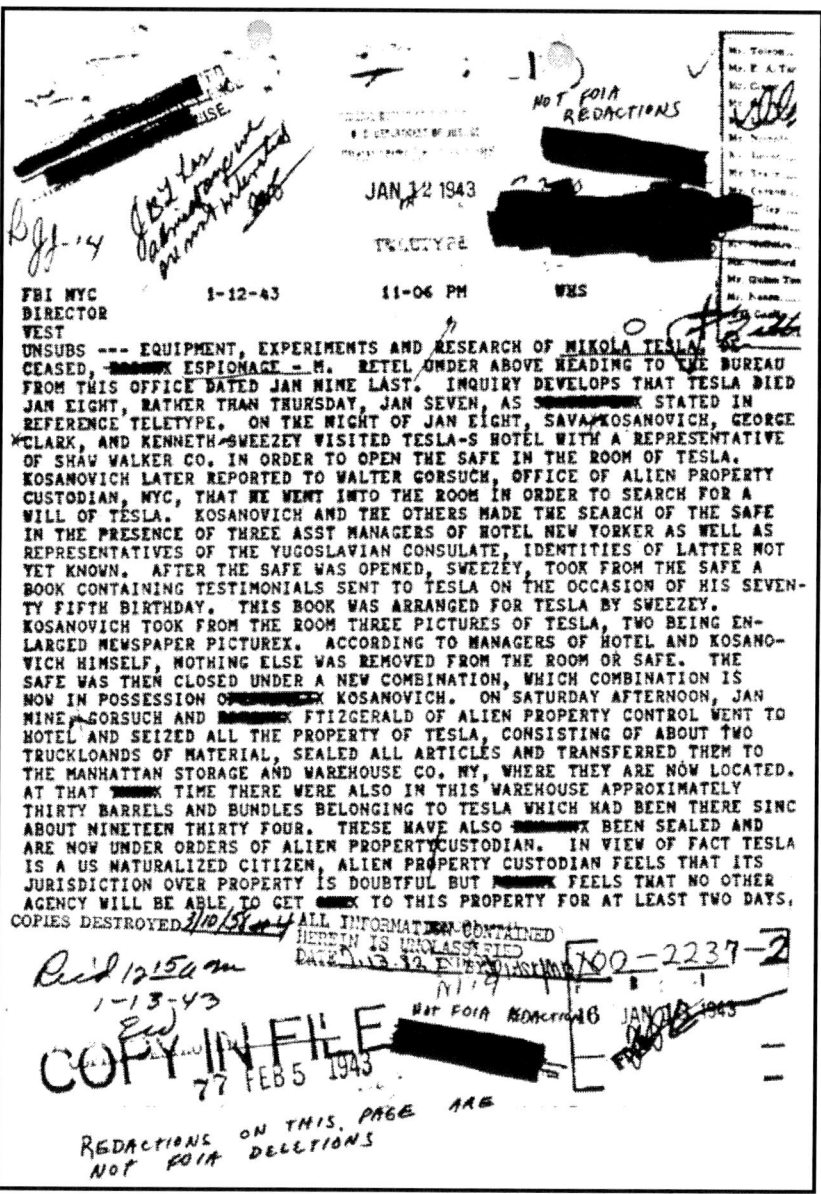

NOT FOIA REDACTIONS

JAN 12 1943

TELETYPE

FBI NYC   1-12-43   11-06 PM   WHS
DIRECTOR
WEST
UNSUBS --- EQUIPMENT, EXPERIMENTS AND RESEARCH OF NIKOLA TESLA, DE
CEASED, ESPIONAGE - M. RETEL UNDER ABOVE HEADING TO THE BUREAU
FROM THIS OFFICE DATED JAN NINE LAST. INQUIRY DEVELOPS THAT TESLA DIED
JAN EIGHT, RATHER THAN THURSDAY, JAN SEVEN, AS STATED IN
REFERENCE TELETYPE. ON THE NIGHT OF JAN EIGHT, SAVA KOSANOVICH, GEORGE
CLARK, AND KENNETH SWEEZEY VISITED TESLA-S HOTEL WITH A REPRESENTATIVE
OF SHAW WALKER CO. IN ORDER TO OPEN THE SAFE IN THE ROOM OF TESLA.
KOSANOVICH LATER REPORTED TO WALTER GORSUCH, OFFICE OF ALIEN PROPERTY
CUSTODIAN, NYC, THAT HE WENT INTO THE ROOM IN ORDER TO SEARCH FOR A
WILL OF TESLA. KOSANOVICH AND THE OTHERS MADE THE SEARCH OF THE SAFE
IN THE PRESENCE OF THREE ASST MANAGERS OF HOTEL NEW YORKER AS WELL AS
REPRESENTATIVES OF THE YUGOSLAVIAN CONSULATE, IDENTITIES OF LATTER NOT
YET KNOWN. AFTER THE SAFE WAS OPENED, SWEEZEY, TOOK FROM THE SAFE A
BOOK CONTAINING TESTIMONIALS SENT TO TESLA ON THE OCCASION OF HIS SEVEN-
TY FIFTH BIRTHDAY. THIS BOOK WAS ARRANGED FOR TESLA BY SWEEZEY.
KOSANOVICH TOOK FROM THE ROOM THREE PICTURES OF TESLA, TWO BEING EN-
LARGED NEWSPAPER PICTUREX. ACCORDING TO MANAGERS OF HOTEL AND KOSANO-
VICH HIMSELF, NOTHING ELSE WAS REMOVED FROM THE ROOM OR SAFE. THE
SAFE WAS THEN CLOSED UNDER A NEW COMBINATION, WHICH COMBINATION IS
NOW IN POSSESSION OF KOSANOVICH. ON SATURDAY AFTERNOON, JAN
NINE, GORSUCH AND FTIZGERALD OF ALIEN PROPERTY CONTROL WENT TO
HOTEL AND SEIZED ALL THE PROPERTY OF TESLA, CONSISTING OF ABOUT TWO
TRUCKLOANDS OF MATERIAL, SEALED ALL ARTICLES AND TRANSFERRED THEM TO
THE MANHATTAN STORAGE AND WAREHOUSE CO. NY, WHERE THEY ARE NOW LOCATED.
AT THAT TIME THERE WERE ALSO IN THIS WAREHOUSE APPROXIMATELY
THIRTY BARRELS AND BUNDLES BELONGING TO TESLA WHICH HAD BEEN THERE SINC
ABOUT NINETEEN THIRTY FOUR. THESE HAVE ALSO BEEN SEALED AND
ARE NOW UNDER ORDERS OF ALIEN PROPERTY CUSTODIAN. IN VIEW OF FACT TESLA
IS A US NATURALIZED CITIZEN, ALIEN PROPERTY CUSTODIAN FEELS THAT ITS
JURISDICTION OVER PROPERTY IS DOUBTFUL BUT FEELS THAT NO OTHER
AGENCY WILL BE ABLE TO GET TO THIS PROPERTY FOR AT LEAST TWO DAYS,
COPIES DESTROYED 3/10/51   ALL INFORMATION CONTAINED
HEREIN IS UNCLASSIFIED
DATE 3 12   BY

100-2237-2

NOT FOIA REDACTIONS

COPY IN FILE
77 FEB 5 1943

REDACTIONS ON THIS PAGE ARE
NOT FOIA DELETIONS

305

Enlarged: Document 2

First half of the page

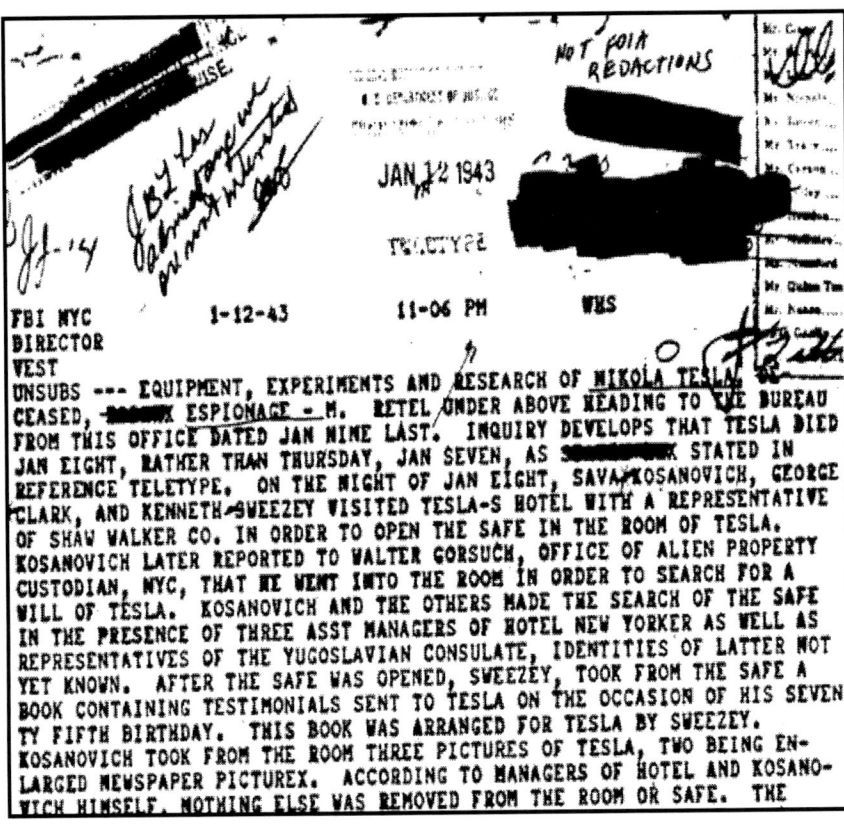

NOT FOIA REDACTIONS

JAN 12 1943

TELETYPE

FBI NYC   1-12-43   11-06 PM   WMS
DIRECTOR
WEST
UNSUBS --- EQUIPMENT, EXPERIMENTS AND RESEARCH OF NIKOLA TESLA, DE-
CEASED, ▮▮▮▮▮ ESPIONAGE - M. RETEL UNDER ABOVE HEADING TO THE BUREAU
FROM THIS OFFICE DATED JAN NINE LAST. INQUIRY DEVELOPS THAT TESLA DIED
JAN EIGHT, RATHER THAN THURSDAY, JAN SEVEN, AS ▮▮▮▮▮ STATED IN
REFERENCE TELETYPE. ON THE NIGHT OF JAN EIGHT, SAVA KOSANOVICH, GEORGE
CLARK, AND KENNETH SWEEZEY VISITED TESLA-S HOTEL WITH A REPRESENTATIVE
OF SHAW WALKER CO. IN ORDER TO OPEN THE SAFE IN THE ROOM OF TESLA.
KOSANOVICH LATER REPORTED TO WALTER GORSUCH, OFFICE OF ALIEN PROPERTY
CUSTODIAN, NYC, THAT HE WENT INTO THE ROOM IN ORDER TO SEARCH FOR A
WILL OF TESLA. KOSANOVICH AND THE OTHERS MADE THE SEARCH OF THE SAFE
IN THE PRESENCE OF THREE ASST MANAGERS OF HOTEL NEW YORKER AS WELL AS
REPRESENTATIVES OF THE YUGOSLAVIAN CONSULATE, IDENTITIES OF LATTER NOT
YET KNOWN. AFTER THE SAFE WAS OPENED, SWEEZEY, TOOK FROM THE SAFE A
BOOK CONTAINING TESTIMONIALS SENT TO TESLA ON THE OCCASION OF HIS SEVEN-
TY FIFTH BIRTHDAY. THIS BOOK WAS ARRANGED FOR TESLA BY SWEEZEY.
KOSANOVICH TOOK FROM THE ROOM THREE PICTURES OF TESLA, TWO BEING EN-
LARGED NEWSPAPER PICTUREX. ACCORDING TO MANAGERS OF HOTEL AND KOSANO-
VICH HIMSELF, NOTHING ELSE WAS REMOVED FROM THE ROOM OR SAFE. THE

*** *** ***

Enlarged: Document 2

Second half of the same page

VICH HIMSELF; NOTHING ELSE WAS
SAFE WAS THEN CLOSED UNDER A NEW COMBINATION, WHICH COMBINATION IS
NOW IN POSSESSION O███████ KOSANOVICH. ON SATURDAY AFTERNOON, JAN
NINE, GORSUCH AND ████████ FTIZGERALD OF ALIEN PROPERTY CONTROL WENT TO
HOTEL AND SEIZED ALL THE PROPERTY OF TESLA, CONSISTING OF ABOUT TWO
TRUCKLOANDS OF MATERIAL, SEALED ALL ARTICLES AND TRANSFERRED THEM TO
THE MANHATTAN STORAGE AND WAREHOUSE CO. NY, WHERE THEY ARE NOW LOCATED.
AT THAT ████ TIME THERE WERE ALSO IN THIS WAREHOUSE APPROXIMATELY
THIRTY BARRELS AND BUNDLES BELONGING TO TESLA WHICH HAD BEEN THERE SINC
ABOUT NINETEEN THIRTY FOUR. THESE HAVE ALSO ████████ BEEN SEALED AND
ARE NOW UNDER ORDERS OF ALIEN PROPERTY CUSTODIAN. IN VIEW OF FACT TESLA
IS A US NATURALIZED CITIZEN, ALIEN PROPERTY CUSTODIAN FEELS THAT ITS
JURISDICTION OVER PROPERTY IS DOUBTFUL BUT ██████ FEELS THAT NO OTHER
AGENCY WILL BE ABLE TO GET ████ TO THIS PROPERTY FOR AT LEAST TWO DAYS.
COPIES DESTROYED 3/10/5? ALL INFORMATION CONTAINED
HEREIN IS UNCLASSIFIED
DATE 1-13-32 FBI ████████ 100-2237-2

Rec'd 12 15 am
1-13-43

Not FOIA REDACTION 6 JAN 1943

COPY IN FILE

77 FEB 5 1943

REDACTIONS ON THIS PAGE ARE
NOT FOIA DELETIONS

*** *** ***

307

Document 3: Office Memorandum. United States Government

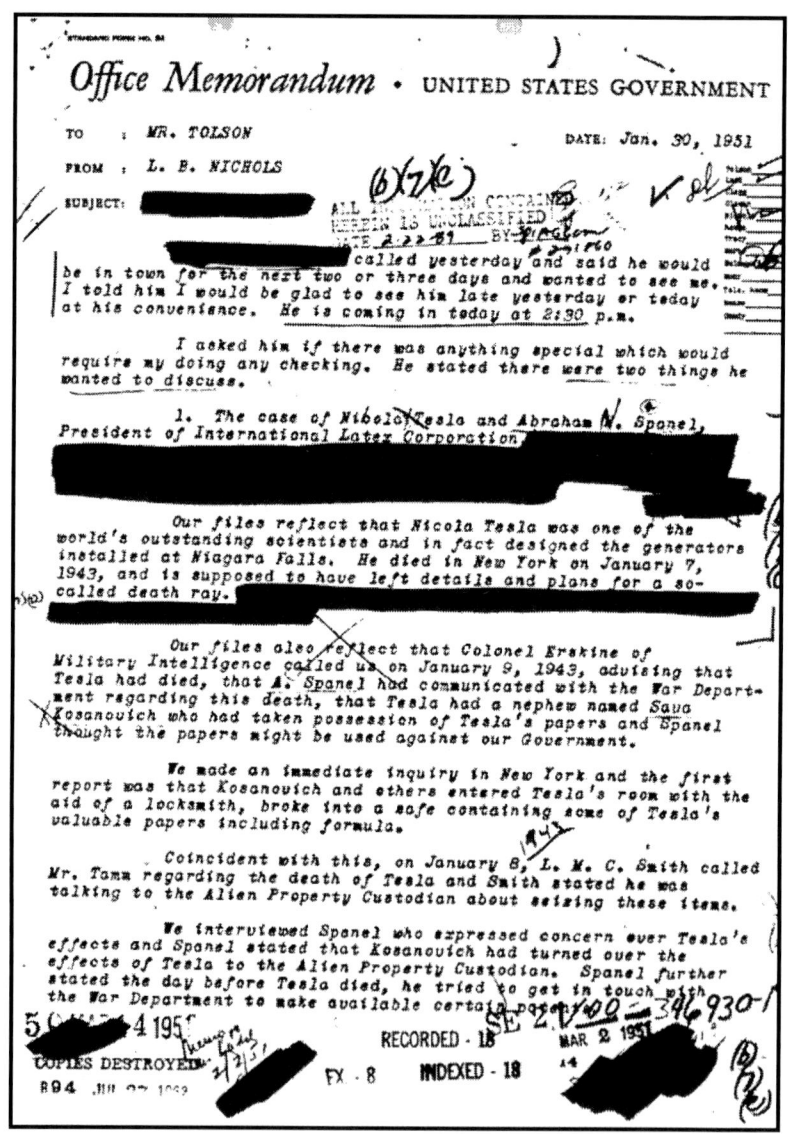

Office Memorandum · UNITED STATES GOVERNMENT

TO : MR. TOLSON   DATE: Jan. 30, 1951

FROM : L. B. NICHOLS

SUBJECT: ▮▮▮▮▮▮▮▮▮▮▮

ALL INFORMATION CONTAINED
HEREIN IS UNCLASSIFIED
DATE 2-22-89 BY ▮▮▮▮

▮▮▮▮ called yesterday and said he would
be in town for the next two or three days and wanted to see me.
I told him I would be glad to see him late yesterday or teday
at his convenience. He is coming in today at 2:30 p.m.

I asked him if there was anything special which would
require my doing any checking. He stated there were two things he
wanted to discuss.

     1. The case of Nicola Tesla and Abraham N. Spanel,
President of International Latex Corporation ▮▮▮▮▮▮▮▮▮▮▮▮▮▮▮▮▮▮▮▮▮▮▮
▮▮▮
▮▮▮

     Our files reflect that Nicola Tesla was one of the
world's outstanding scientists and in fact designed the generators
installed at Niagara Falls. He died in New York on January 7,
1943, and is supposed to have left details and plans for a so-
called death ray. ▮▮▮▮▮▮▮▮▮▮▮▮▮▮▮▮▮▮▮▮▮

     Our files also reflect that Colonel Erskine of
Military Intelligence called us on January 9, 1943, advising that
Tesla had died, that A. Spanel had communicated with the War Depart-
ment regarding this death, that Tesla had a nephew named Sava
Kosanovich who had taken possession of Tesla's papers and Spanel
thought the papers might be used against our Government.

     We made an immediate inquiry in New York and the first
report was that Kosanovich and others entered Tesla's room with the
aid of a locksmith, broke into a safe containing some of Tesla's
valuable papers including formula.

     Coincident with this, on January 8, L. M. C. Smith called
Mr. Tamm regarding the death of Tesla and Smith stated he was
talking to the Alien Property Custodian about seizing these items.

     We interviewed Spanel who expressed concern over Tesla's
effects and Spanel stated that Kosanovich had turned over the
effects of Tesla to the Alien Property Custodian. Spanel further
stated the day before Tesla died, he tried to get in touch with
the War Department to make available certain papers ▮▮▮▮

SE 2 V 00 - 346930 - 1

5 ▮▮▮▮ 4 195▮   RECORDED · 18   MAR 2 1951

COPIES DESTROYED ▮▮▮▮▮   FX · 8   INDEXED · 18

894 JUL 27 1962

308

Enlarged: Document 3; Office Memorandum.
United States Government

First half of the page

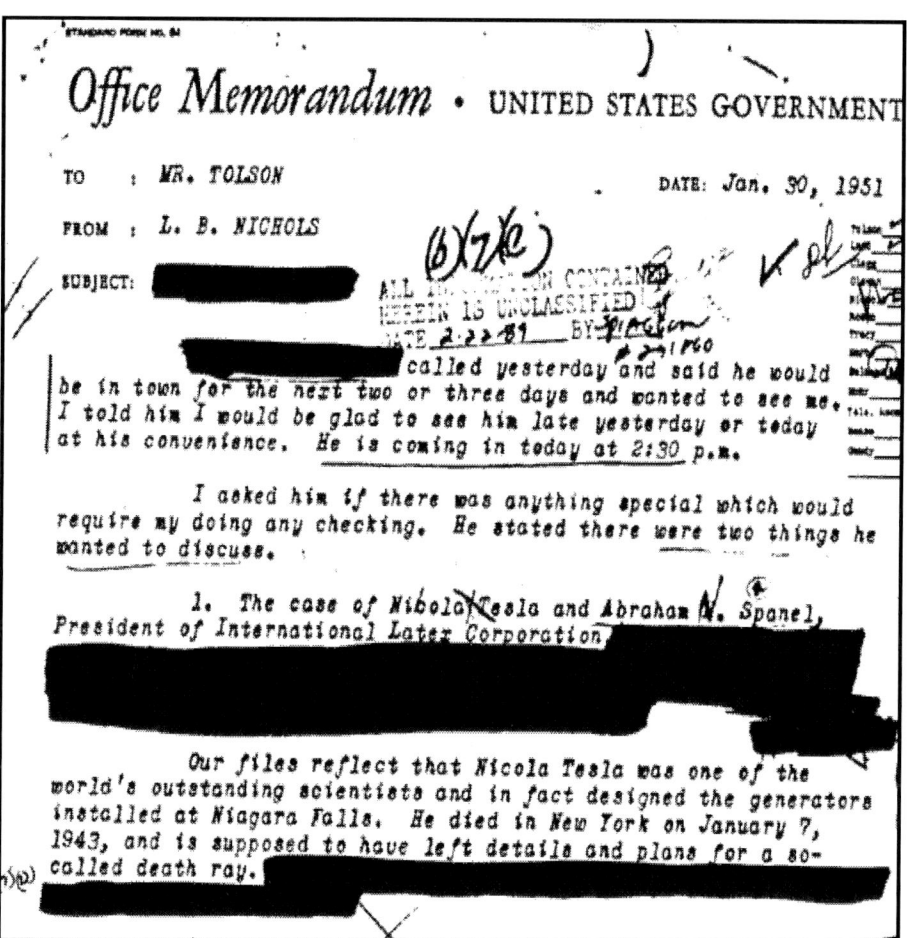

Office Memorandum • UNITED STATES GOVERNMENT

TO : MR. TOLSON   DATE: Jan. 30, 1951

FROM : L. B. NICHOLS

SUBJECT: ▮▮▮▮▮▮▮▮

(b)(7)(c)

ALL INFORMATION CONTAINED
HEREIN IS UNCLASSIFIED
DATE 2-22-81 BY ▮▮▮▮

called yesterday and said he would
be in town for the next two or three days and wanted to see me.
I told him I would be glad to see him late yesterday or today
at his convenience. He is coming in today at 2:30 p.m.

I asked him if there was anything special which would
require my doing any checking. He stated there were two things he
wanted to discuss.

1. The case of Nikola Tesla and Abraham N. Spanel,
President of International Latex Corporation ▮▮▮▮▮▮▮▮▮

▮▮▮▮▮▮▮▮▮▮▮▮▮▮▮▮▮▮▮▮▮▮▮▮▮▮▮▮▮▮▮▮▮▮▮▮▮▮

Our files reflect that Nicola Tesla was one of the
world's outstanding scientists and in fact designed the generators
installed at Niagara Falls. He died in New York on January 7,
1943, and is supposed to have left details and plans for a so-
called death ray. ▮▮▮▮▮▮▮▮▮▮▮▮▮▮▮▮▮▮▮▮

***   ***   ***

Enlarged: Document 3; Office Memorandum. United States Government

Second half of the page

Our files also reflect that Colonel Erskine of Military Intelligence called us on January 9, 1943, advising that Tesla had died, that A. Spanel had communicated with the War Department regarding this death, that Tesla had a nephew named Sava Kosanovich who had taken possession of Tesla's papers and Spanel thought the papers might be used against our Government.

We made an immediate inquiry in New York and the first report was that Kosanovich and others entered Tesla's room with the aid of a locksmith, broke into a safe containing some of Tesla's valuable papers including formula.

Coincident with this, on January 8, L. M. C. Smith called Mr. Tamm regarding the death of Tesla and Smith stated he was talking to the Alien Property Custodian about seizing these items.

We interviewed Spanel who expressed concern over Tesla's effects and Spanel stated that Kosanovich had turned over the effects of Tesla to the Alien Property Custodian. Spanel further stated the day before Tesla died, he tried to get in touch with the War Department to make available certain papers.

RECORDED - 18   MAR 2 1951

COPIES DESTROYED   FX - 8   INDEXED - 18

R94

***   ***   ***

Letter to Mr. Clarence Kelly, Director of, F.B.I.

Signature Censored!!!

April 20, 1976

Mr. Clarence Kelly
Director
F.B.I.
Washington, DC

ALL INFORMATION CONTAINED
HEREIN IS UNCLASSIFIED
DATE 2-3-80 BY [signature]

Dear Mr. Kelly:

Mr. Allen and Mr. Ruchlehaus,   former acting Director of the
FBI, contacted me in 1973 regarding the unavailability of
American microfilm records of Nikola Tesla's unpublished diary
(now in the Belgrade museum, arranged by month per folder).

At the time I discounted the possibility that these unpublished
discoveries had military significance. But because of experiments
now under way at Hill AFB, I now suspect such military
applications exist and feel it imperative that you be notified,
particularly in view of the fact that the Soviets have primary
access to the entire collection.

Two photos of each page exist.

After Tesla's death, scientists from the Navy and OSS performed
a cursory examination of the diary and notes, which if my
memory serves me correctly, was one month long, hardly enough
time to decipher Tesla's torturous handwriting. Though Tesla
wrote in English, his penmanship was small, blurred, and as
difficult to translate as a foreign language.

According to the museum director (1971), the Soviets had made
copies of some portions, but not the Colorado Springs diary,
which numbers 500 pages, 20 that directly pertain to ball
lightning, and 20 or so relevant to the equipment construction.
(We copied the most significant portions, but feel more exists)

FX-115   REC-52 /00 - 2237 -   29

_____ an article, _____ magazine, EDN (an electrical
engineering magazine), but only with the very recent receipt_____
of an unpublished manuscript from John J. O'Neill's book
(PRODIGAL GENIUS) did I place credence on Tesla's later claim
to military applications. Incidentally, some of O'Neill's
descriptions were inaccurate and exagerated, as we have exceeded
Tesla's results and are familiar with the experiments. At any
rate, there are three possible military applications.

311

First, Tesla claimed that the lightning balls (which destroyed his equipment) could be used to destroy aircraft. I have talked to AF personnel --such as ████████████████████ ██████████ who saw one inside his plane in flight--and found AF personnel fear these "rf balls," as they call them.

Second, it is a suspicion of mine that ball lightning, if injected with lithium, could produce a cheap fusion bomb.

Third--and this may be no more than a suspicion--the propulsion mode of ball lightning involves electro-gravitic interaction, by which means air vehicles of revolutionary configuration may be constructed. There are no presently-known laws of physics that can account for the propulsion (400 mph or so when following an airliner). Other hitherto unsuspected applications may exist.

None of these applications were the goal of Project Tesla, which centered on producing ball lightning as Tesla did and studying it as a plasma confinement technique for fusion reactors. Incidentally, Tesla's claim to setting up standing waves on the earth's surface (wireless power) was erroneous and involved techniques similar to Project Sanguine, that is, using the earth's atmosphere as a waveguide (█████████ is aware of our research).

Cordially,

P.S. By a copy of this letter, along with the enclosures, I am notifying the C.I.A.

Enclosures: 2

*** *** ***

Letter from the Department of Defense to the F.B.I.

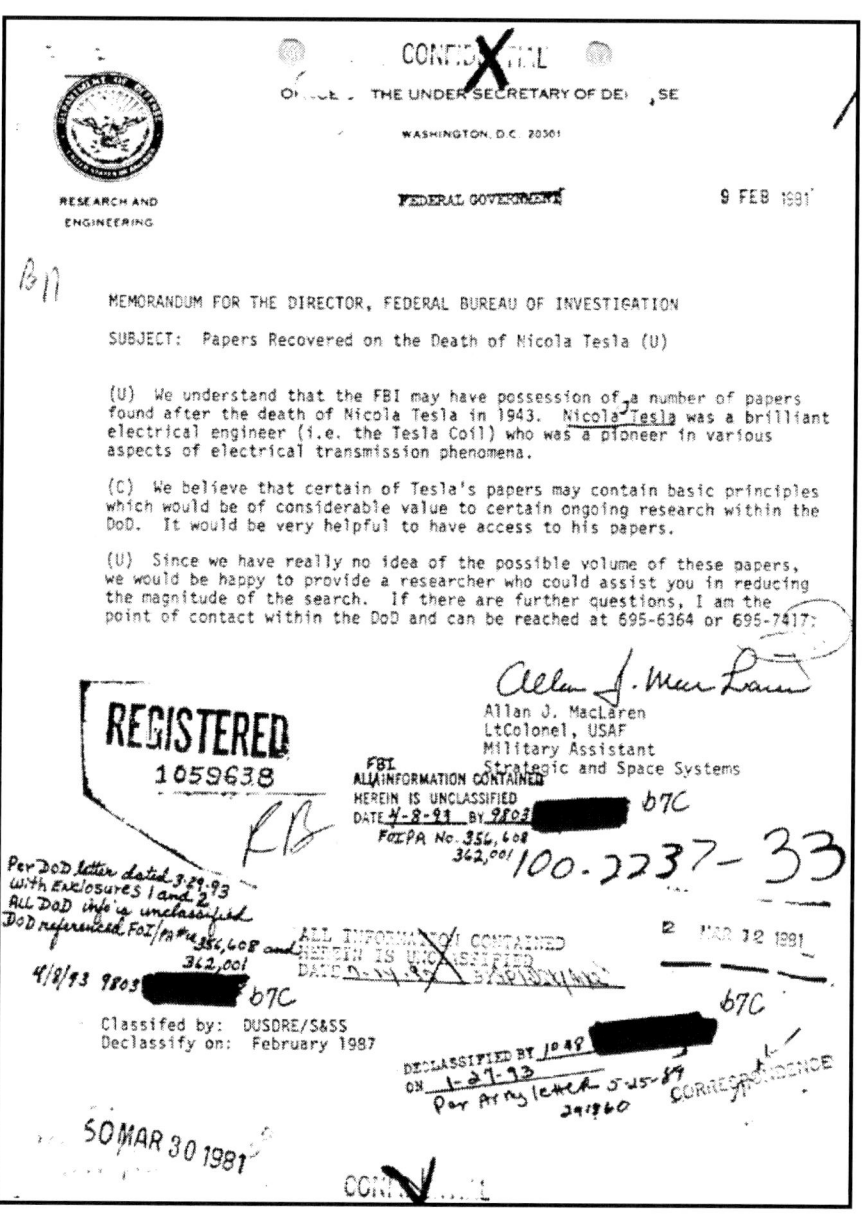

CONFIDENTIAL

OFFICE OF THE UNDER SECRETARY OF DEFENSE

WASHINGTON, D.C. 20301

RESEARCH AND
ENGINEERING

FEDERAL GOVERNMENT

9 FEB 1981

MEMORANDUM FOR THE DIRECTOR, FEDERAL BUREAU OF INVESTIGATION

SUBJECT: Papers Recovered on the Death of Nicola Tesla (U)

(U) We understand that the FBI may have possession of a number of papers found after the death of Nicola Tesla in 1943. Nicola Tesla was a brilliant electrical engineer (i.e. the Tesla Coil) who was a pioneer in various aspects of electrical transmission phenomena.

(C) We believe that certain of Tesla's papers may contain basic principles which would be of considerable value to certain ongoing research within the DoD. It would be very helpful to have access to his papers.

(U) Since we have really no idea of the possible volume of these papers, we would be happy to provide a researcher who could assist you in reducing the magnitude of the search. If there are further questions, I am the point of contact within the DoD and can be reached at 695-6364 or 695-7417.

Allan J. MacLaren
LtColonel, USAF
Military Assistant
Strategic and Space Systems

REGISTERED
1059638

FBI
ALL INFORMATION CONTAINED
HEREIN IS UNCLASSIFIED
DATE 4-8-93 BY 9803   b7C
FOIPA No. 356,608
362,001   100.2237-33

Per DoD letter dated 3-29-93
with Enclosures 1 and 2
ALL DoD info is unclassified
DoD referenced FoI/pa #356,608 and
362,001
4/8/93 9803   b7C

ALL INFORMATION CONTAINED
HEREIN IS UNCLASSIFIED
DATE   BY

2   MAR 12 1981

b7C

Classifed by: DUSDRE/S&SS
Declassify on: February 1987

DECLASSIFIED BY 1048
ON 1-27-93
Per Army letter 5-05-89
291960   CORRESPONDENCE

50 MAR 30 1981

CONFIDENTIAL

313

Even though, this document is heavily censored, explosive information can be deduced from its contents. Pay attention to the lines censored before "at Wright-Patterson Air Force", and before "at Dayton, Ohio." I made references to these two locations in my article.

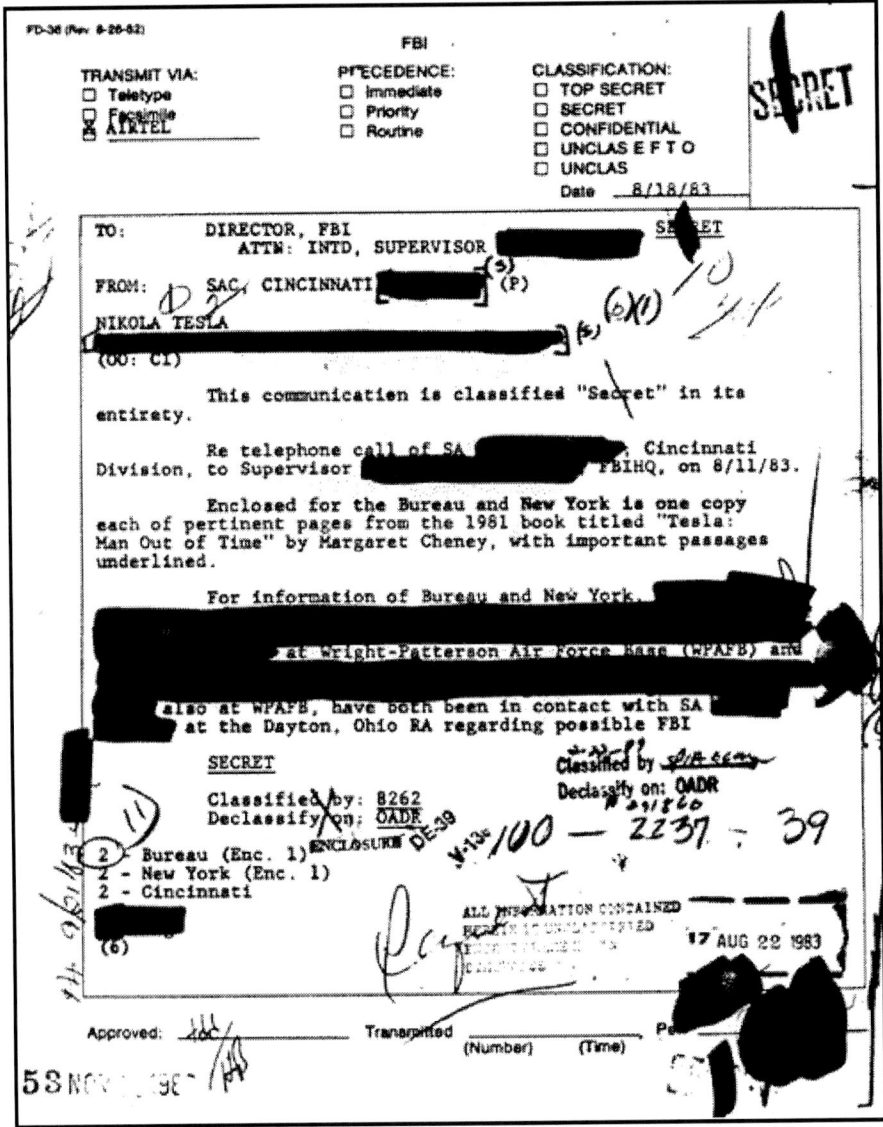

involvement in the seizing of Nikola Tesla's research papers and other documents and scientific instruments after his death on January 7, 1943.   1956

They both explained that Tesla was a scientific genius and experimenter who was born in Yugoslavia of Serbian parents on 7/10/56, went to school later in Gratz, Austria, Prague, Czechoslovakia and Paris, France. He immigrated to the U.S. in the early 1880's, worked for Thomas Edison's laboratory for a couple of years, then started his own lab after being paid $1 million dollars for rights to his patents on his polyphase systems of alternating current dynamos, which lead to the harnessing of Niagra Falls for producing electricity and then the power system of the whole country. He was naturalized in 1889. He predicted wireless communication (radio). His later experiments in Colorado and elsewhere lead to his producing artificial lightning in the millions of volts. He also had patents on the concept of neon and flourescent lights, but he later made little money on his later inventions, although he continued to do experiments leading to devices of great potential worth, which he never patented. He became more reclusive in his later years, living in various hotels in New York City. In the 1930's he claimed he had developed the concept and method of building a "death ray", which could destroy planes at many miles distant, for defending America. Also, there are reports of resonance machines or devices whereby he could shake one or many large city buildings from some distance away.

Both [████] and [████] said that Tesla donated "some" of his papers (or copies thereof) to the Tesla Institute in Belgrade, Yugoslavia; set up in the 1930's in his honor by their government. Biographies on Tesla claim that either the custodian of Alien Property and/or the FBI seized his papers and other personal effects, including a safe or safes, and other property immediately after his death in 1943. This is elaborated on in the enclosed copies of certain pages of Margaret Cheney's book, "Tesla: Man Out of Time".

[████] said that after World War II Tesla's papers were shipped to the Tesla Institute in Belgrade, Yugoslavia, by his nephew, Sava Kosanovic, who had become Tito's Ambassador to the U.S. There were reports that some microfilming of Tesla's papers by government agents while they were still in storage in New York under Kosanovic's custody.

Also, the Soviet Union has allegedly had access
to some of Tesla's papers, possibly in Belgrade and/or
else where, which influenced their early research into directed
energy weapons, and ████ feels access to much of Tesla's
papers on lightning, beam weapons and/or "death rays" would
give him more insight into the Soviet beam weapons program.
This is Butler's area of expertise and responsibility. He
has been unable to locate any Tesla papers or copies of same
in the classified or unclassified libraries at WPAFB. However,
there are reports that some portions of them were shipped by
the Custodian of Alien Property Office in Washington, D.C. to
a technical research lab at WPAFB, possibly the "Equipment
Lab", now closed for some years or reorganized into another
organization.

████ and ████ are both desirous of learning
the locations of such papers of Tesla as now exist in the U.S.
for both intelligence and research purposes. Therefore, ████
would like to examine FBI files relating to Nikola Tesla and
possibly any on Sava Kosanovic, his nephew who received the
bulk of his papers after Tesla's death, and may possibly
have been the subject of FBI investigation.

████ travels to the Washington, D.C. area on
FTD business periodically and can review FBI files at FBIHQ
relating to Tesla and Kosanovic.

REQUEST OF THE BUREAU

The Bureau is requested to conduct full indices checks
on both Nikola Tesla and Sava Kosanovic.

Should there be such files at FBIHQ, as well as at
New York, it is requested that Bureau consider granting the
above ████ of FTD, official access to same, in the
interest of national security.

LEADS

NEW YORK

AT NEW YORK, NEW YORK

Will conduct same indices check as requested of
Bureau and advise the Bureau and Cincinnati of results and
confirm such files and references still exist there.

-3-　　　　　　　　　　　　　　　SECRET

Note: Von Braun and Dr. Teller notes on the aliens' threats and rapture (Previously at the Los Alamos secret archives), now in the hands of the United States Air Force, Strategic Command, NSA, and the "Vaults" of the CIA.

*** *** ***